12/2/75

CONSERVING ENERGY

CONSERVING ENERGY

Prospects and Opportunities
in the New York Region

JOEL DARMSTADTER

Published for Resources for the Future, Inc.
by The Johns Hopkins University Press
Baltimore and London

This book is a product of RFF's Division of Energy and Resource Commodities, directed by Hans H. Landsberg. It was edited by Joan R. Tron.

RFF editors: Herbert C. Morton, Joan R. Tron, Ruth B. Haas, Jo Hinkel

The Johns Hopkins University Press, Baltimore, Maryland 21218
The Johns Hopkins University Press Ltd., London

Library of Congress Catalog Card Number 75-15414
ISBN 0-8018-1775-7 cloth
ISBN 0-8018-1776-5 paper

Library of Congress Cataloging in Publication data will be found
on the last printed page of this book.

Contents

TABLES

FIGURE

Preface

WHAT IS the long-term outlook for dampening the growth of energy consumption in the New York area? This study throws some cautiously optimistic, yet realistic, light on this question. In his appraisal, the author takes account of existing patterns of energy use and regional economic structure; and, with those elements in mind, he looks at energy conservation possibilities in the light of market adjustments, institutional changes, and selected policy initiatives. Even though New York City and its multicounty environs possess numerous unique demographic, economic, and energy using characteristics, I believe that many of the author's observations provide insight into the problem of achieving energy conservation that have wider geographic—indeed, national —relevance

In the issues addressed in this book, which, for the most part, was researched prior to, and written during, the swirling national and international energy upheavals of 1973–74, Joel Darmstadter sets his sights along a longer-range time path rather than trying to grapple with problems of overriding current anxiety. But where recent events seem to mark a turning point of enduring significance, he has taken these into consideration.

The study is the concluding publication in a series of three reports[1] dealing with energy use in the New York area. The project as a whole, supported by a special grant from the Ford Foundation, involved a collaborative research effort between Resources for the Future and the Regional Plan Association, New York City.

Hans H. Landsberg
*Director, Division of Energy
and Resource Commodities*

December 1974

[1] Two earlier releases in this series were: Regional Plan Association and Resources for the Future, *Patterns of Energy Consumption in the Greater New York City Area: A Statistical Compendium*, RFF Working Paper EN-2 (Washington: Resources for the Future, July 1973). Regional Plan Association and Resources for the Future, *Regional Energy Consumption* (RPA, New York, 1974).

Acknowledgments

THANKS are due to reviewers of an earlier version of this study. Their comments proved most helpful in the final revision. The reviewers included Michael Gerrard, Council on the Environment of New York City; John H. Gibbons, University of Tennessee; Olaf Hausgaard, State of New York Public Service Commission; Joseph T. Kirdahy, Jr., Consolidated Edison Co.; Sam H. Schurr, Electric Power Research Institute; and Robert T. Taussig, Columbia University. Within RFF, valuable criticism came from Irving Hoch, Hans H. Landsberg, Paul R. Portney, James W. Sawyer, John J. Schanz, Jr., and Frederick J. Wells. Edward Ames, of the Ford Foundation, and Sam H. Schurr (then at RFF), provided helpful guidance at the initiation of the project as a whole.

At the Regional Plan Association, Boris Pushkarev (Vice President) and Regina Armstrong (Chief Economist) also contributed notably to the entire effort.

In addition, I am grateful to Elizabeth K. Vogely for her meticulous research assistance during the greater part of the project; to Joan R. Tron for editorial help; and to Helen-Marie Streich for her usual secretarial excellence.

Introduction

ENERGY IS desired by consumers in a variety of forms—electricity, gaseous and liquid fuels, and (to a limited extent) coal. In its various forms, energy serves a broad variety of purposes: the heating, cooling, and lighting of homes and nonresidential buildings; the mechanization of industrial processes; the fueling of private and public transport; and others. Throughout most of the past century, an abundant store of domestic energy resources meant that these and other needs could be accommodated without major difficulty. U.S. energy prices were low and, compared with increases in the price of other things, held steady over prolonged periods of time, thereby reducing the intermittent anxiety over resource depletion. Dependence on foreign sources was minimal, and the environmental burden of producing and consuming large and steadily growing amounts of fuels and power was, at least until recent decades, barely perceived. Fundamental and successive transformations of the nation's energy base—first from fuelwood to coal, then from coal to petroleum and gas—could be relatively smoothly absorbed into a dynamic industrial society. In brief, whatever else may at times have marred the performance of the American economic system, access to cheap, plentiful, and secure supplies of energy could normally be taken for granted.

That idyllic state of affairs has now vanished. Events during the past decade, capped by the traumatic jolt of the Middle East oil crisis, have irretrievably confronted this nation with a strikingly new perspective within which to view and manage its affairs in the energy field. A variety of factors have brought us to this juncture. Throughout the past decade, nationwide energy consumption grew more rapidly than did the combined domestic output of conventional fuels and nuclear energy. The leveling-off in U.S. output of oil and gas (in the wake of declining exploration and reserve development) and environmental restrictions limiting the use of coal led to a need for greatly expanded imports of oil—the balancing energy source in times of stringency. The exercise of effective monopoly control by the major oil-exporting countries, resulting in precipitous price increases for world oil as well as in uncertainty about future supply prospects, has enormously aggravated the nation's energy dilemma.

Inevitably, some of these problems have had a disproportionately severe impact on certain regions of the United States. The New York metropolitan area, along with the northeastern part of the country in general, has experienced the energy pinch and increased costs in a particularly marked way. Electric utilities in New York City and its surroundings have been, for some years now, almost entirely dependent on foreign residual fuel oil for their oil-fired boilers. The latter, in turn, are the area's predominant electric generating mode, coal having been essentially phased out due to air pollution control regulations, and atomic power plant sites still encountering the resistance of environmental groups and others made uneasy by the nearby presence of nuclear stations. Indeed, given the dense agglomeration of people in the New York area and the dwindling availability of recreational land in the periphery of the city (e.g., Long Island Sound and the Hudson River shoreline) large power generating facilities of all kinds are increasingly unwelcome intruders throughout much of the area. Even before the Middle East oil embargo and steeply increased world oil prices exerted their additionally harsh impact on the local region, this area had long been subjected to higher fuel and power costs than those prevailing in the country as a whole. The area is, of course, located far from presently worked centers of U.S. oil and gas resources. Moreover, prolonged resistance to deep water ports, offshore Atlantic coast petroleum development, and construction of new refinery capacity could further aggravate the situation.

The things we have listed here all relate to actual or potential barriers to an expanded *supply* of energy for the region; or, at least, to a supply of energy available at acceptable conditions of cost and reliability. Until very recently, persons having to deal nationally or regionally with such issues—electric utility planners, equipment manufacturers, oil company executives, regulatory commissioners—would have been unlikely to question the assumptions underlying the estimated growth of future energy *demand* to which expansion of energy supply was supposed to accommodate itself. In a climate of cheap and easily expandable domestic energy supply, its use uninhibited by environmental restrictions, that viewpoint did not seem especially objectionable. With the recognition that a large-scale accretion to U.S. energy availability may in fact be a quite formidable task in terms of tolerable criteria of cost, environment, and security, a questioning of this traditional acquiescence to exponential demand growth—indeed, a deeper questioning of the ethical postulates underlying unremitting growth in energy consumption—has aroused the interest of many persons throughout the country. The questioning ranges from those in sparsely settled regions of the United States who want to preserve the perceived amenities of an uncluttered environment to those who, as in the New York area, seek desperately to juggle what they see as a precarious balance between growth and livability.

All of which is not meant to prejudge the question of whether reasonable energy objectives towards which we may aspire in the next decade or so can more purposefully be approached by an attempt to hold down the rate of growth in energy consumption than by an expansion of supply. As a matter of fact, it seems pretty clear even on casual reading that substantially greater domestic capacity to produce oil, natural gas, coal and nuclear power, and such "secondary" energy products as gasoline and thermal electricity, will be necessary in the years ahead. But it is a measure of the new thinking that even federal authorities, normally given to an expansive economic perspective, are coupling their preoccupation with policies to expand energy supplies with a clearcut emphasis on factors which may succeed in dampening the future increase in energy consumption.

Our report is devoted to the second problem—an evaluation of the potential for more sparing energy utilization. The New York region was selected to illustrate conservation possibilities for reasons that derive from earlier remarks. The area's range of energy supply options is tightly constrained. That is, there is heavy dependence on uncertain fuel oil imports; energy prices are higher than elsewhere, and, in the case of electricity, seemingly inadequate to finance operations of a major utility system; and planning for expansion of electric operating capability is an awesome process subject to lengthy adjudication over environmentally related issues. Construction of a 2 million kilowatt pumped storage plant at Cornwall, first proposed by Consolidated Edison in 1962, was still bogged down in judicial dispute in 1974. For these reasons, the possibility of a serious imbalance between present and prospective electricity supply and demand appears to be particularly acute. But disruptive shortfalls of other energy forms, especially natural gas, are also feared.

The New York region is thus not only an interesting, but perhaps a valuable, testing ground for a systematic evaluation of energy conservation potentialities, in spite of having some energy characteristics not shared by other regions of the country. Understanding and advance recognition of which energy use savings are possible with relative ease and which might be achieved only with great difficulty could help avert the more drastic, arbitrary, and inequitable cutbacks which a severe supply crisis might bring about—either by means of prohibitive price increases or through governmental control measures. But to repeat what has already been stressed: to argue the case for conservation in energy use is not to cast doubt on the importance of expansion in energy supply; it should only redress to a degree the balance between these joint and complementary aspects of rational energy management.

The geographic, conceptual, and statistical underpinnings for this report vary with the analytical focus. Because they are the locus of the more severe problems mentioned above, particular geographic stress is at times put on the five boroughs (counties) of New York City and the three proximate counties of Westchester, Nassau, and Suffolk. We term this eight-county grouping the "Greater New York City Area." For broader analytical and descriptive purposes, this eight-county grouping is too restrictive. For example, insight into energy usage associated with transportation patterns is much more adequately captured if a wider geographic span is considered. In this report, we adopt as our most comprehensive geographic measure, the Regional Plan Association's thirty-one-county "region," embracing the above eight counties plus six additional counties in New York, three counties in Connecticut, and fourteen counties in New Jersey (see Figure 1). Thus, depending on the particular issue being highlighted (but depending also on the availability of data), our references may be to New York City, the Greater NYC Area, or the thirty-one-county region (hereafter referred to as the NYR). However, some ideas or issues are discussed outside any specific geographic framework at all. These are points that cut across regional bounds.

Our treatment of energy forms as they are ultimately delivered to final customers also tends to be somewhat pragmatic. Because regional demand–supply imbalances have been especially keenly felt in the provision of electric power, electric utilities, and particularly their deliveries to residential users, will occasionally be singled out for special emphasis. But because of the interrelatedness of different energy resources—oil, gas, coal, uranium, and falling water are all alternative sources for producing electricity, while electricity, gas, and oil are substitutable for each other in space heating—the entire range of fuels and power components must be dealt with as an aggregate.

For the most part, the historical period covered by this report spans the decennial benchmark years 1960 and 1970, since these years span the interval for which numerous Census and other data are available for purposes of constructing our most comprehensive energy measures. In certain cases, however, we have gone back to pre-1960 figures, as well as recorded year-to-year figures during 1960–70 where that was possible and advisable. In addition, recognizing the onrush of developments in the energy field nationally and regionally, we have tried to take account of events since 1970 so as to portray as realistic a "base-year" case as possible.

While this report is not primarily designed to be an effort in projecting future developments, it is pertinent, using conditional forecasts for the years ahead, to illustrate the consequences of some hypothetical conservation practices and measures in energy use. For it is obviously insufficient to hypothesize the implications of, say, a specified improvement in insulation standards, without also speculating about the time period during which such a change could be brought about. For this purpose, we have selected the "intermediate term" to the mid-1980s as our target period, although here again, both nearer-term prospects as well as longer-range possibilities are at times taken into account.

To convey some notion of where significant potentials for energy saving may be, it is useful to begin by noting the principal constituents of national and regional energy demand and their rates of increase. This is done in Chapter 1. Chapter 2 considers how regional energy consumption, and its major components, might grow over the next decade

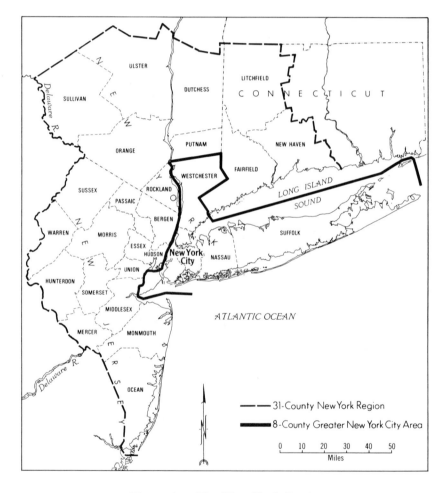

Figure 1.   The New York Region

or so without purposeful measures to dampen demand. This baseline case enables us to judge the relative contribution to "unconstrained" demand of a variety of assumed conservation practices and measures, as detailed in chapters 3 and 4. Chapter 5 compares prospective demand growth with and without explicit conservation efforts. The sixth chapter presents some concluding remarks.

1.

Recent Patterns of
Energy Consumption

EVEN WITHOUT quantitative substantiation, one can intuitively point to some of the key factors influencing a region's demand for energy. Chief among these would be the size of the area's population, work force, and income; regional fuel and power costs; and, because industries and services vary widely in their degree of energy use, the "mix" of regional economic activities. Population density appears to be another important feature influencing the amount of energy consumed. And climate plays an obvious, if in the aggregate often less conspicuous, role. There is, of course, a certain interdependence within such an array; a region's particular industrial structure, for example, clearly reflects in part the presence or absence of low-cost energy.

In the light of this range of determining factors, it is scarcely surprising that different regions of the United States exhibit marked variations in their energy-use characteristics, both at a point in time and over a period of years. Thus, if the principal difference between, say, Louisiana and Pennsylvania were merely in their respective per capita income levels, Pennsylvania—outranking Louisiana in this respect by around 35 percent—could be expected to be the greater per capita energy consumer. The fact that Louisiana's per capita energy consumption actually exceeds Pennsylvania's by over 100 percent is due in large measure to the concentration of energy-intensive industries attracted to the relatively low fuel and power costs prevailing in this important energy-producing area of the country.[1]

[1] State-by-state energy consumption data appear in U.S. Department of the Interior, *United States Energy Fact Sheets by States and Regions* (Washington, February 1973).

So it is that New York City and its tristate surroundings exhibit a number of energy-using characteristics bearing the distinct imprint of particular locational economic and demographic circumstances. It may be instructive, at the outset of this report, to highlight some of these features by means of some broad indicators for the thirty-one-county tristate region (NYR), its narrower geographic subdivisions, and the United States as a whole. These data, presented in tables 1-1, 1-2 and 1-3, are adopted, for the most part, from the document, *Regional Energy Consumption* (hereafter referred to as *REC*)—a prior release of the project of which the present study forms the concluding report.[2] Indeed, much of the quantitative underpinning for the present study comes from that previous publication, which will therefore be cited recurrently throughout these pages.

Major Consuming Sectors

Compared with nationwide consumption patterns, the most pronounced characteristics of NYR energy usage evident in these tables are three-fold: a relatively low level of per capita utilization of fuels and power, a markedly different distribution than for the nation as a whole in the major components of energy usage, and a record of growth in energy consumption which in recent years has been conspicuously below the national trend line. The first two of these three factors are directly related, insofar as the principal reason for the region's relatively low level of overall net energy consumption[3] in terms of population size—and in spite of what is by national standards a high level of income—is the disproportionately low level of energy demand in the industrial and transportation sectors of its economy. This is most readily evident from the last panel of Table 1-1, where we see that per capita industrial use of energy in the NYR is less than 17 percent of the national per

[2] *Regional Energy Consumption*, Second Interim Report of a Joint Study by Regional Plan Association, Inc. (RPA) and Resources for the Future, Inc. (RPA, New York, 1974). The study contains voluminous county-by-county detail for the 31-county RPA region. A still earlier and shorter joint RPA-RFF report dealt primarily with the Greater NYC Area, although it presented selected national comparisons as well. That first interim report was entitled *Patterns of Energy Consumption in the Greater New York City Area: A Statistical Compendium* (released by RFF as Working Paper EN-2, Washington, D.C., 1973).

[3] Terminology is described in the notes to Table 1-1.

capita average, and transportation energy demand less than 71 percent. And these low region-to-nation relatives are not close to being offset by the relatively high per capita levels of NYR energy demand in the residential and commercial sectors. Reflecting these contrasting patterns of usage, 38 percent of national energy consumption is concentrated in the industrial sector of the economy, while less than 10 percent is so deployed in the NYR. (Even in the NYR outside the Greater NYC Area, which includes the nearby New Jersey manufacturing and refinery centers, the industrial share is only 14 percent.) Conversely, 30 percent of U.S. energy consumption goes to residential, commercial, and public facility uses, while 57 percent of NYR consumption reflects these sectoral end-uses.

This contrast, which is still further heightened when New York City or the Greater NYC Area are compared with the nation, does not, of course, reveal a particularly surprising phenomenon. All it shows is that the New York region, particularly its subareas, is neither attractive to the location of manufacturing activity which is significantly dependent on fuels and power inputs—e.g., metallurgy—nor, given a dense settlement pattern served by a reasonably widespread public transport system, is the role of the private passenger car, with its high energy requirements, anywhere near as preeminent there as it is elsewhere in the country.

What has been said of energy consumption as a whole is as true for electricity alone; here, again, per capita consumption levels are significantly below the national average. And, as with net energy consumption in the aggregate, it is the disproportionately low level of the NYR's industrial electric power consumption which is the accountable factor. It is interesting to note, however (and for reasons that are not clear), that, in what industry *does* reside in the NYR, electricity constitutes a greater relative share of industrial energy consumed than is the case nationally (see the top panel of Table 1-3). This is also the case in the transport sector, owing to the disproportionate importance of electrified transit and railways. Per capita electricity use in the NYR's residential sector is also markedly lower than nationally. Apartment living—so widespread in the region—is an important reason; it limits the variety of electric uses possible in individual homes. But, as with the nation as a whole, electricity has been accounting for rising shares of total NYR energy consumption.

Table 1-1. Regional and National Energy Consumption and Related Indicators, 1960 and 1970

Indicators	New York City (1)	West-chester (2)	Nassau-Suffolk (3)	Greater NYC Area 1970 (4)	Rest of NYR (5)	NYR total (31 counties) [a] (6)	United States (7)	NYR as percent of United States 1970 (8)	NYR as percent of United States 1960 (9)
Population (thousands)	7,896	894	2,556	11,346	8,410	19,756	203,212	9.7	9.8
Money income (million 1969 dollars)	29,373	4,542	10,461	44,376	31,936	76,312	635,563	12.0	12.6
Net energy consumption (trillion Btu)									
Total	984	168	469	1,622	1,494	3,398	51,894	6.6	7.0
Residential	420	62	171	652	419	1,071	9,548	11.2	12.5
Gross energy consumption (trillion Btu)	1,291	196	549	2,036	1,857	4,186	63,571	6.6	7.0
Electricity consumption (thousand Mwh)	28,837	3,525	9,811	42,172	43,741	85,914	1,402,988	6.1	6.2
Per capita figures									
Money income (1969 dollars)	3,720	5,078	4,093	3,911	3,797	3,863	3,118	123.9	128.6
Net energy consumption (million Btu)									
Total	124.6	187.9	183.7	142.9	177.7	172.0	255.3	67.4	71.3
Residential	53.1	69.1	67.0	57.5	49.8	54.2	46.9	115.6	127.6
Gross energy consumption (million Btu)	163.5	219.6	214.7	179.5	220.8	211.9	312.8	67.7	71.2
Electricity consumption (kwh)	3,652	3,943	3,838	3,717	5,201	4,349	6,884	63.2	63.3
Per dollar of money income figures									
Net energy consumption (million Btu)									
Total	33.5	36.9	44.9	36.5	46.8	44.5	81.7	54.5	55.4
Residential	14.3	13.7	16.3	14.7	13.1	14.0	15.0	93.3	98.9
Gross energy consumption (million Btu)	44.0	43.2	52.5	45.9	58.1	54.9	100.0	54.9	55.3
Electricity consumption (kwh)	0.98	0.78	0.94	0.95	1.37	1.13	2.21	51.1	48.8

Percent of net energy consumption
by sector:[b]

Residential	42.6	36.8	36.5	40.2	28.0	31.5	18.4	11.2	12.5
Commercial/public facilities	31.4	28.3	27.7	30.0	25.6	25.6	11.8	14.2	14.3
Industrial	6.4	6.4	4.2	5.8	14.3	9.3	37.7	1.6	1.9
Transportation	19.5	28.5	31.6	23.9	32.0	33.6	32.1	6.9	6.7

Net per capita energy consumption
by sector (million Btu)

Residential	53.1	69.1	67.0	57.5	49.8	54.2	46.9	115.6	127.6
Commercial/public facilities	39.2	53.1	51.0	42.9	45.5	44.0	30.1	146.2	145.7
Industrial	8.0	12.0	7.7	8.3	25.5	16.0	96.1	16.6	19.5
Transportation	24.3	53.5	58.0	34.2	56.9	57.7	81.9	70.4	68.4

Sources: Taken from the two **RPA-RFF** joint interim reports published earlier in the course of this project or from sources cited in those two studies. The more recent of these was *Regional Energy Consumption* (hereafter referred to as *REC*), second interim report of a Joint Study by Regional Plan Association, Inc. and Resources for the Future, Inc. (released by RPA, New York City, 1974). "The first interim report was *Patterns of Energy Consumption in the Greater New York City Area: A Statistical Compendium* (released by RFF as RFF Working Paper EN-2, Washington, D.C., 1973). In particular, we relied upon tables A, B, 17, and 19 of the second interim report; and the summary tables (and notes thereto) in the first interim report. In a few cases, the present table embodies slight revisions in the figures appearing in the interim reports.

Notes:

Money income, not to be confused with the related concept of "personal income," refers to receipts by families and unrelated individuals, by place of residence, as derived from the U.S. decennial Census of Population. In the panel of the table relating energy and electricity consumption to dollars of income, the ratios for subregional areas are distorted by the fact that the area's energy use is related to income earned by residents of the area rather than income generated by the area's economy (including commuters). The ratios are nonetheless shown since, corrected for the above distortion, areas such as NYC would show even lower ratios of energy consumption relative to income. Less distortion applies to the residential energy use–income ratio and to the thirty-one-county average for all four ratios shown.

Net energy consumption comprises fuels (coal, oil, gas) delivered to the residential, commercial/public facility, industrial, and transportation sectors along with steam and electric utility delivered to each of these sectors. (Electricity, net of transmission losses, is converted at the direct heat equivalent of 3,412 Btu [British thermal units] per kwh.)

Gross energy consumption subtracts utility steam and electricity deliveries from net energy consumption but then adds in fuels used at utility plants for the production of steam and generation ef electricity.

In both the net and gross energy consumption measures, energy sources used as nonfuel raw material inputs in the economy (around 5 percent of gross energy consumption nationwide), such as oil feedstocks for the petrochemical industry, have been deducted from the national totals so as to permit comparison with available regional data excluding this component. (Lubricating oil for vehicles is, however, included both regionally and nationally.)

[a] Regional totals or averages may differ from sum of subregional areas because they include geographically unallocable items not shown separately.

[b] In the national shares, there is included a small miscellaneous category, not shown separately. Also in this panel, the region-to-U.S. relatives refer to the underlying absolutes rather than to the percentage shares shown.

Not only is the NYR's energy use low by the national norm, its rate of increase has also been quite modest, as Table 1-2 shows for the period 1960–70. This has been true in spite of NYR population and real income growth rates that are only moderately less than the nationwide rates. (For NYC alone, however, population levels were practically stationary during the decade; there, net energy consumption per capita increased at an average annual rate of 1 percent while estimated energy usage in the residential sector actually declined.) Contrary to overall energy growth, NYR electric power growth lagged only slightly behind national growth; NYC and the Greater NYC Area trailed the nationwide electricity growth rate by a more noticeable margin, but still recorded substantially rising electricity demand.

While, as the foregoing illustrates, NYR energy consumption levels and growth both tend to be disproportionately lower than the comparative national figures, that finding may have no more than curiosity value to an area which, in many respects, faces energy problems that transcend similar problems at the national level. For example, with the dense population pattern that characterizes the NYR, even relatively low levels of energy use signify intensely concentrated combustion emissions on an areawide basis, thus posing potentially more acute environmental hazards than in a more dispersed population setting. In the Greater NYC Area alone, energy consumption per square mile of area in 1970 was approximately sixty times the ratio prevailing nationally and thirteen times that for New York State as a whole.[4]

Regional Energy Resource Requirements

The region's relatively low level of per capita energy use should not obscure the fact that its mix of raw energy requirements predisposes the NYR—and particularly the city—to be relatively more vulnerable to uncertainty in foreign supply sources and domestic fuel shortages.

---

[4] See RFF, *Patterns of Energy Consumption in the Greater New York City Area*, Table C. An interesting energy-related aspect of density has been analyzed by Irving Hoch of Resources for the Future. He found that winter temperature, for given latitude, increases significantly with density, but that summer temperature is much less affected. This "heat island" effect could be a factor reducing NYC energy consumption below levels otherwise prevailing and could be a net benefit, somewhat tending to offset pollution costs. (Hoch surmises that the heat island winter benefits outweigh summer costs.)

Table 1-2.   Average Annual Rates of Change in Regional and National Energy Consumption and Related Indicators, 1960–1970

Indicator	New York City	Greater NYC Area	Region (31 counties)	United States
Population	0.1	0.7	1.2	1.3
Money income				
(based on 1969 dollars)	3.0	3.6	4.1	4.6
Net energy consumption:				
Total	1.1	2.0	2.8	3.5
Residential	−0.3	0.5	1.3	2.4
Gross energy consumption	1.9	2.6	3.4	4.1
Electricity consumption	5.2	6.2	7.2	7.4
Per capita:				
Money income	2.8	2.8	3.2	3.3
Net energy consumption:				
Total	1.0	1.2	1.6	2.2
Residential	−0.4	−0.2	0.1	1.1
Gross energy consumption	1.7	1.9	2.3	2.8
Electricity consumption	5.1	5.5	6.0	6.0

For notes and sources, see Table 1-1.

The NYR's economy has a pattern of energy resource dependence which differs rather distinctly from the national picture. The contrast emerges clearly in Table 1-4. Over 75 percent of the region's energy resource inputs are liquid fuels, while the comparable national share is about 42 percent. The situation in the electric power sector alone is particularly striking. There, oil constituted 62 percent of regional fuel inputs in 1970 compared with just 13 percent for the United States as a whole.

This pattern within the overall fuel mix arises from the disproportionately greater role that coal continues to play on the national scene compared with its much more modest NYR importance. Again, the electric power sector highlights the situation most vividly. In 1960, coal constituted half the generating station fuel inputs, both regionally and nationally. By 1970, largely under the impact of air quality control regulations (as well as greater access to quota-free imported residual fuel oil), NYR power-station coal use was down to 18 percent, while in the country as a whole it still stood at around 45 percent. Since 1970, the NYR's reliance on coal has further slipped, especially in the Greater NYC Area where central generating stations have all but abandoned

Table 1-3. Regional and National Electricity Consumption, by Sector, 1960 and 1970

Sector	New York City	West-chester	Nassau-Suffolk 1970	Greater NYC Area	Rest of region	Region (31 counties) 1970	Region (31 counties) 1960	United States 1970	United States 1960
Percentage share of electricity in sectoral energy consumption:									
Residential	6.8	7.9	8.8	7.4	11.9	9.2	4.6	16.0	9.4
Commercial and public facilities	14.1	10.8	10.3	12.7	13.3	13.0	7.8	19.7	9.4
Industrial	26.8	16.4	24.3	25.1	22.4	22.7	16.1	10.4	9.0
Transportation	4.9	0.6	0.2	2.6	0.1	0.9	1.4	0.1	0.2
All sectors	10.0	7.2	7.1	8.9	10.0	8.6	5.6	9.2	6.3
Percentage distribution, by sector, of total electricity consumption:									
Residential	29.1	40.3	44.8	33.7	33.3	33.5	29.5	31.9	29.7
Commercial and public facilities	44.2	42.5	40.1	43.1	34.0	38.5	35.7	25.1	19.1
Industrial	17.2	14.7	14.2	16.3	32.2	24.4	28.7	42.7	50.5
Transportation	9.5	2.5	0.9	6.9	0.5	3.6	6.7	0.3	0.7
All sectors	100.0	100.0	100.0	100.0	100.0	100.0	100.0	100.0	100.0
Per capita electricity consumption, by sector (million Btu direct equivalent):[a]									
Residential	3.6	5.4	5.9	4.3	5.9	5.0	2.4	7.5	3.9
Commercial and public facilities	5.5	5.7	5.3	5.5	6.0	5.7	3.0	5.9	2.5
Industrial	2.1	2.0	1.9	2.1	5.7	3.6	2.3	10.0	6.6
Transportation	1.2	0.3	0.1	0.9	0.1	0.5	0.6	0.1	0.1
All sectors	12.5	13.5	13.1	12.7	17.7	14.8	8.3	23.5	13.1

For notes and sources, see Table 1-1.
[a] See Appendix for method of converting kwh to Btu.

coal as a boiler fuel. Of course, the use of coal for electric power is now receding nationally as well, though it remains to be seen whether the drive toward a greater degree of national energy self-sufficiency (and, in the longer run, success with coal gasification and liquefaction) will stem that trend.

In the recent past, nuclear power contributed only a minor portion of NYR and national electricity generation (hence a still less significant fraction of gross energy consumption). However, not only is nuclear power's role quite important in particular utility systems, such as Con Edison's, but its proportionate importance is now increasing rapidly, both locally and nationally.

End-Use Patterns in Greater Detail

To address the question of trends and conservation potentials in different components of end-use energy consumption it will be necessary to go beyond the rather aggregative sectoral consumption clusters introduced above (residential, commercial and public facilities, etc.) and deal with a more specific set of energy-use categories.

Unfortunately, the effort to develop this kind of quantitative specificity has occurred more at the national level than regionally. However, both by (hopefully) judicious inferences from national data and use of fragmentary NYR information, it should prove possible to approach levels of detail helpful to the overall analysis. Table 1-5 shows a reasonably exhaustive tabulation of recent net energy consumption in the United States by end-use categories. It becomes quickly apparent that in each of the major sectors of the economy, one or two items dominate the entire sector: space heating in the residential as well as commercial sector; process steam and direct heat applications in industry; automobiles and trucking in the transport sector. In fact, the foregoing small number of identifiable end uses of energy account for 77 percent of national energy consumption.[5] It is noteworthy that water heating, air conditioning, refrigeration, cooking, and lighting account for only an additional 9 percent, bringing the figure up to 85 percent of the total.

[5] Excluding nonfuel and power applications of energy commodities, as explained in Table 1-1.

Table 1-4.   Gross Energy Resource Inputs, Thirty-one-County New York Region and United States, 1960 and 1970

percentage distribution

Inputs	31-county region 1960	31-county region 1970	United States 1960	United States 1970
Coal				
Power stations	9.5	4.9	9.6	11.7
Other uses	5.6	1.2	14.0	8.3
Natural gas				
Power stations	3.0	3.7	4.1	6.3
Other uses	8.6	12.3	24.2	27.1
Oil				
Power stations	5.4	16.6	1.3	3.3
Other uses	66.0	59.6	42.8	38.8
Nuclear[a] and hydro power and electric power imports	1.1	1.6	4.1	4.5
Total	100.0	100.0	100.0	100.0
In trillion Btu	2,988	4,186	43,649	63,571
Energy resource inputs at electric power stations				
Coal	50.1	18.1	50.3	45.5
Natural gas	15.6	14.0	21.5	24.4
Oil	28.6	62.0	6.8	12.7
Nuclear[a] and hydro power and electric power imports	5.7	5.9	21.4	17.4
Total	100.0	100.0	100.0	100.0
In trillion Btu	546	1,081	8,317	16,458

Sources: Regional data from *REC,* primarily Table E. National data for 1960 from U.S. Department of the Interior, *An Energy Model for the United States,* Information Circular 8384, Washington, 1968; for 1970 from U.S. Department of the Interior, *Minerals Yearbook 1971,* Washington, 1973.

Note: Power station fuel use refers to electricity generation as well as steam production.

[a] There was no nuclear power in the region in 1960.

It is also interesting to observe how automotive energy use swamps air travel (passenger and freight combined) in relative importance.

To be sure, certain end uses—particularly air conditioning and air transport—have tended in recent years to record the fastest annual growth rates among energy uses, even though still representing only modest portions of annual nationwide energy consumption. There has, in addition, been rapid growth in the relative contribution of electricity to the different types of end-use consumption shown in Table 1-5 and to energy use in the aggregate, as noted earlier. (The table itself does not break out electricity separately, however.) This trend, notable, for example, in residential space heating, has generated a good deal of public debate, at times highly contentious. Critics point out that electricity is a relatively inefficient way of converting primary energy resources to useful energy and should not be resorted to in uses, such as space heating, which can be accomplished by other means. In addition, they point to the increasingly difficult task of siting power plants under conditions that are judged environmentally benign. It might nevertheless be noted that while electricity's contribution to energy consumption in space heating is clearly on a continuing uptrend, its current share in household space heating is probably still not much above 5 percent nationally, while New York State has been estimated as having the lowest percentage "saturation" of electric heat of any state in the nation.[6] And, as with other ostensibly "inefficient" energy-use practices, if electric heat is priced to reflect its social costs, and if policy (e.g., national energy independence) does not compel demand restraint, people willing to absorb the cost of luxuriating in electric heat ought perhaps to have the freedom to do so. (The "ifs" in the preceding sentence are, however, major provisos —too often blithely employed to assume problems away. We revert to such questions later in the study.)

We have yet to judge more fully the extent to which NYR end-use consumption patterns conform to the national picture. Table 1-5 clearly indicates which end-use components have the greatest quantitative significance nationally. It is a fair presumption that the key national items stand out as major components in the NYR as well, and thus ought to be evaluated as to possible energy conservation potentials regionally. Indeed, from scattered available data, for the NYR, it appears that savings opportunities in space heating and transport would seem to warrant especially close scrutiny. Thus, one calculation for NYC alone (the Brookhaven estimate cited in footnote 6) suggests that space heating in the residential and commercial/public facilities sectors add up to nearly half the city's net energy consumption (the respective sectoral shares were 27 and 20 percent). Household and commercial uses, which are much farther down the line in their proportionate contribution to

---

[6] New York State Interdepartmental Fuel and Energy Committee, *Final Report of the Ad Hoc Committee On Appliance and Apparatus Efficiency*, June 1973, p. 54. For NYC alone, electricity's share in residential space heating is reckoned as still insignificant. See Brookhaven National Laboratory estimate shown in Council on the Environment of New York City, *Energy and the New York City Environment* [New York, 1974], pp. 58–59. The matter of its *growth* is, however, a hotly contested issue.

Table 1-5.   Net Energy Consumption in the United States, by Sector and Estimated End Use, 1972

Sector and end use	Consumption (trillion Btu)	Percent distribution
Residential	10,454	18.9
Space heat	6,826	12.3
Water heat	1,522	2.7
Air conditioning	296	0.5
Refrigeration	288	0.5
Lighting	200	0.4
Cooking	470	0.8
Other	852	1.5
Commercial	6,560	11.8
Space heat	4,431	8.0
Water heat	527	1.0
Air conditioning	540	1.0
Refrigeration	260	0.5
Lighting	300	0.5
Cooking	148	0.3
Other	354	0.6
Industrial	20,293	36.7
Process steam	11,456	20.7
Direct heat	6,110	11.0
Electric drive	1,929	3.5
Electrolytic	402	0.7
Other	396	0.7
Transportation	18,054	32.6
Auto	10,071	18.2
Truck	3,978	7.2
Rail	463	0.8
Air	1,407	2.5
Other	2,135	3.9
U.S. net energy consumption	55,361	100.0

Source: Federal Power Commission, National Power Survey, *Forecasts of Electric Energy and Demand to the Year 2000,* A Report by the Task Force on Forecast Review to the Technical Advisory Committee on Power Supply (mimeographed report dated August 24, 1973, Exhibit 6). This report, in turn, represents an updated version of estimates for 1968 prepared by Stanford Research Institute, *Patterns of Energy Consumption in the United States,* A Report to the Office of Science and Technology, Washington, 1972. Estimated energy use for residential and commercial lighting seemed low; the figures were scaled up on the basis of estimates prepared at the Oak Ridge National Laboratory and the Ford Foundation Energy Policy Project.

Note: In conformity with our practice in Table 1-1, nonfuel uses have been subtracted from the tabulation in the source cited above.

NYC's net energy use, are residential water heating (6 percent) and air conditioning in both sectors combined (3½ percent).

One can surmise that the proportionate importance of space heating in the region's twenty-six counties outside NYC would be somewhat less (conversely: other uses relatively more important) because of the greater incidence of air conditioners and other energy-using appliances in the outlying areas. For example, ownership saturation is distinctly higher for the non-NYC part of the region than for the city in central or multiple room air conditioning units, dishwashers, and multiple TV sets.[7] But notwithstanding intraregional variations in its relative role, space heating is clearly of key importance in the NYR's energy use.

The next most important item is private automotive travel, whose share of net energy consumption comes to 18 percent of the regional total. Energy consumption by trucks and by airline traffic (both passenger and freight) amount to another 6 percent each; for airline traffic, the estimate is based on fuel consumption of aircraft departing from the NYR's airports.[8] Although the indicated automotive energy share of 18 percent is similar to the national percentage (Table 1-5), it represents far less use on a per capita basis, as indicated earlier. Moreover, within the NYR, automotive energy use is dramatically less prominent in NYC than elsewhere, as is clearly suggested by the "transportation" line in the next-to-last panel of Table 1-1.

Of course, it is not sufficient to confine our energy conservation assessment solely to those consumption categories which we have identified as predominant now. Components currently of relatively small overall importance—e.g., air conditioning and electric space heating—could, at prevailing rates of growth, come to assume far more significance in subsequent years. We are interested in evaluating not only the potentials for greater energy husbandry in existing practices but also the outlook for blunting the steepness of future growth. In addition, important aspects of energy conservation are left out of account in the rather broad annual tallies which we have recorded here. One example—of

[7] See *Regional Energy Consumption*, Table 23.

[8] Ibid., Table F. Because of this limitation, the share of regional energy consumption accounted for by the transportation sector reflects a somewhat synthetic statistical construct. There is clearly some distortion if primary interest in the data centers on where the fuel is actually consumed. (In this case, it could be over the mid-Atlantic or the Great Lakes.) If, on the other hand, one is concerned with the regional economy's fuel supply adequacy—as was the case, for example, during the 1973-74 winter jet fuel shortage—then the measurement practice used here is much more valid. (In a sense, an analogous statistical limitation applies to our treatment of road transport, but since proportionately more of the gasoline or diesel fuel bought within the region is expended in the region compared with long-haul aircraft, the result is less distorting.)

critical importance—is the peak demand problem. This is illustrated by the contribution of air conditioning loads to summer peak electricity demand, which for the Con Edison service area are estimated at about 40 percent.[9] The peak demand problem is thus seen to be a potentially crucial factor in electricity supply capability, irrespective of the modest role of air conditioning within the area's overall energy-use pattern. The extent to which consumers might be impelled to shift particular energy demands from peak to offpeak portions of the load curve thus becomes one of the issues in fashioning a conservation strategy, even though what may be involved is a change in *spacing* energy use rather than in *curtailing* it.

Summary

Compared with the nation as a whole, NYR per capita energy use is relatively low—primarily because of the absence of significant quantities of industrial energy demand. (Such energy gets "embodied" in goods produced elsewhere in the country and shipped into the region.) Levels of per capita residential energy consumption in the region are more nearly like the national average; in the commercial sector, the regional figure is decisively higher; while in transport, regional per capita use is below the national norm. In a number of respects, the New York City portion of the region exhibits characteristics unique to its own particular circumstances. Per capita energy use in transportation is well below the regional level and farther still below the U.S. average—a feature arising from limited automotive ownership and travel, combined with reasonable access to public transport. Energy consumption growth in the region has tended to lag behind nationwide energy growth—a lag that is only in part the consequence of a regionally slower pace of population and income growth. However, in the case of electricity—as opposed to total energy—past regional growth has closely paralleled high national rates of increase. Space heating and transportation, accounting for significant shares of energy use, deserve critical examination from a conservation perspective, as does air conditioning, the demand for which is rapidly growing.

---

[9] Information from Consolidated Edison.

2.

Projections of
Energy Consumption
Without Conservation

IN THIS chapter, we attempt to estimate how the New York region's consumption of fuels and power might change between now and the mid-1980s, in the absence of deliberate energy conservation practices and policies. In essence, this means extrapolating forward what would appear to be the more persistent past trends—trends which are themselves assumed not to reflect any inherent momentum towards the realization of major energy savings. Such a projection can then be used (in subsequent chapters) as a reference case against which we can gauge the quantitative impact of specified conservation actions.

The Nature of the Projections

The approach used here is admittedly easier to verbalize than to set forth operationally. It is not at all a straightforward matter to select an unambiguous energy-trend extrapolation that realistically satisfies the "baseline" conditions necessary for the conservation analysis. That is, even a trend projection may reflect factors already operating to curb growth in energy consumption. Here, the historical period chosen as the basis for the projection becomes important. As an extreme example, a base period which would assign major weight to the 1973–74 period, during which energy prices were rising sharply and demand conspicuously held back, would scarcely serve to define the standard against which explicit conservation actions could be measured. Even a longer

23

historical period, embodying such elements as deceleration in the rate of population increase and what some perceive to be a declining relationship between electricity growth and income growth, may contain inherent tendencies towards dampened energy utilization. Also, policy measures already adopted or mandated for future adoption—for example, upgraded FHA insulation requirements, urban transportation plans to comply with air quality controls or energy efficiency labeling— can, in a sense, be viewed as integral to trends already under way.

To exclude such factors from a trend projection is to bias projected growth on the high side and, therefore, to set up a straw man, insofar as it then enables one to demonstrate that potential payoffs of energy conservation would be large. On the other hand, to modify a trend projection by building in as many demand-dampening factors as we sense are already in progress or "waiting in the wings" would defeat the purpose of the exercise, which is to quantify the effect of stipulated conservation actions on bending the future demand curve.

In the present context, we deem it preferable, as far as practicable, to project future developments as if they were unaffected by explicit demand-dampening tendencies. In practice, this means that we do incorporate the more enduring factors which may tend to depress the volume of energy use—for example, the slowdown in population growth. We do not build in the more recent departures from longer-run trends, such as the precipitous rise in electric power costs, potent though they may be as energy-saving stimuli. If this forces us to err somewhat on the high side of projected NYR energy use, it does have the virtue of conforming to the conventional, "business-as-usual," premises which still govern the thinking of many people.

Only in the course of 1974 did the growth expectations (and corresponding capacity expansion plans) of electric utilities begin to be very much moderated; and, even then, it isn't clear how much of the retrenchment was due to the capital squeeze and how much it was a response to users' conservation impulses. The refrain of decadal doubling of nationwide electric power use, though muted in the context of the 1973–74 slowdown in electricity demand growth, lingers. While such forecasts are not self-fulfilling, neither can they be considered as having no influence on attitudes and decisions about the future. In forecasts by electrical utilities especially, capacity once committed could impel the adoption of ways to promote its maximum use. Viewed in that perspective, it seems to us defensible to accept, at least initially, a view of developments that largely abstracts from the kinds of energy-conserving use patterns that may in fact be already unfolding.

A Look at the Figures

The accompanying set of tables contains projected trends resulting from the approach described above. Three tables (2-1 through 2-3) present data on the underlying demographic and economic factors to which, on the basis of past relationships, projected energy use can broadly, but also reasonably, be linked. These basic quantitative underpinnings

Table 2-1.   Regional Population and Money Income, 1950–1985

	New York City	West-chester County	Nassau-Suffolk	Greater NYC Area	Rest of region	Total region
Population (thousand)						
1950	7,892	626	949	9,467	5,680	15,147
1960	7,782	809	1,967	10,558	7,067	17,624
1970	7,896	894	2,556	11,346	8,410	19,756
1985	7,840	1,050	3,180	12,070	10,445	22,515
Money income (billion current dollars)[a]						
1950	11.63	1.33	1.54	14.50	8.22	22.72
1960	17.94	2.63	5.05	25.62	16.38	42.01
1970	29.37	4.54	10.46	44.38	31.94	76.31
Money income (billion constant 1969 dollars)[a]						
1950	17.58	2.01	2.33	21.91	12.43	34.34
1960	21.87	3.21	6.15	31.24	19.98	51.21
1970	29.37	4.54	10.46	44.38	31.94	76.31
1985	48.02	8.12	21.67	77.81	66.24	144.05
Per capita money income (constant 1969 dollars)[a]						
1950	2,227	3,211	2,452	2,315	2,188	2.267
1960	2,811	3,966	3,129	2,958	2,827	2,906
1970	3,720	5,078	4,093	3,911	3,797	3,863
1985	6,125	7,730	6,814	6,446	6,342	6,397

Notes and Sources: Historical data and projections (the latter prepared by RPA) are taken from Table 19, *REC*, which, along with the text accompanying the present study, may be consulted for additional notes. See also Table 1-1, above, regarding the "money income" concept. Conversion of current to constant dollars was done using the personal consumption deflator in the U.S. Department of Commerce GNP accounts.

[a] Money income figures shown for 1950, 1960, and 1970 (and derived from U.S. Census of Population for those years) refer, respectively, to income actually earned during 1949, 1959, and 1969.

Table 2-2. Regional Data on Households and Housing, 1950–1985

	New York City	West-chester County	Nassau-Suffolk	Greater NYC Area	Rest of region	Total region
Households (millions)						
1950	2.36	0.18	0.26	2.80	1.61	4.41
1960	2.65	0.24	0.52	3.42	2.11	5.52
1970	2.84	0.28	0.70	3.82	2.59	6.41
1985	3.12	0.40	0.97	4.48	3.66	8.14
Average annual housing starts (thousands)[a]						
1950	29.8	6.5	26.2	62.5	40.5	103.0
1960	36.9	4.9	17.5	59.3	56.6	115.9
1970	22.4	3.4	14.8	40.6	53.4	94.0
1985	64.9	12.1	28.3	105.3	124.7	230.1
Dwelling units (millions)						
1950	2.43	0.19	0.32	2.94	1.77	4.71
1960	2.76	0.25	0.59	3.60	2.35	5.95
1970	2.92	0.29	0.75	3.96	2.80	6.76
1985	3.29	0.41	1.07	4.77	4.13	8.90

Source: Taken from Table 21, *REC*, which cites sources for the historical data as well as for the assumptions and methodology underlying the projections.

[a] The four years shown refer, respectively, to average annual housing starts for the periods 1950–59, 1960–69, 1970–72, and 1973–84.

include trends in population and households, residential housing and commercial floorspace (the latter an important energy determining factor in the key office building sector and other commercial enterprises), employment, and money income. Tables 2-4 and 2-5 contain the basic energy-consumption projections. Some general remarks on the substance and derivation of the projections follow. A more specific methodological and analytical discussion appears in a separate document.[1]

It may be useful to refer back briefly to the main elements in the NYR's energy growth in 1960–70. During that decade, total net energy consumption went up 31 percent, averaging 2.8 percent yearly; gross consumption went up some 40 percent, or 3.4 percent yearly.[2]

[1] *Regional Energy Consumption*, Second Interim Report of a Joint Study by Regional Plan Association, Inc. (RPA) and Resources for the Future, Inc. (RPA, New York, N.Y., 1974).

[2] The difference between "gross" and "net" energy consumption arises from the treatment of the electric power and the steam producing sectors. Gross energy counts the fuel inputs needed for power and steam production; net energy counts the quantity of electricity and steam output which, owing to conversion losses, is calorifically substantially below fuel inputs. Measurement of other energy sectors is unaffected by these conventions. See notes to Table 1-1 and the Appendix.

Table 2-3.   Regional Employment and Floorspace, 1950–1985

	New York City	West-chester County	Nassau-Suffolk	Greater NYC Area	Rest of region	Total region
Employment, total (thousands)						
1950	3,909	200	265	4,373	2,292	6,665
1960	3,908	262	542	4,712	2,622	7,333
1970	4,194	338	813	5,345	3,279	8,624
1985	4,462	448	1,100	6,010	4,298	10,308
Employment, manufacturing (thousands)						
1950	992	42	51	1,085	969	2,054
1960	863	58	135	1,056	985	2,042
1970	737	57	165	959	1,008	1,966
1985	592	57	179	828	1,052	1,879
Commercial floorspace[a] (mill. ft^2)						
1960	481	42	89	612	445	1,057
1970	581	67	140	788	595	1,383
1985	665	97	200	962	845	1,808
Total nonresidential floorspace[b] (mill. ft^2)						
1963	1,244	152	327	1,722	1,647	3,369
1970	1,335	160	393	1,888	2,030	3,918

Source: Taken from Table 20, *REC*, which provides notes and references as to derivation and RPA projections.

[a] Commercial floorspace includes retail, services, and most office buildings.

[b] The nonresidential category consists of commercial floorspace plus manufacturing, warehousing, transportation, communications, utilities, and public buildings.

It is possible, approximately, to "decompose" these increments into the main contributory elements. (Note that this does not purport to be an analysis of the *causes* of change, but merely an arithmetic sorting out of what actually took place.) The following tabulation shows the percentage contribution of these elements to the growth in power consumption in 1960–70. As can be seen, the relative importance of the different elements varies somewhat, depending on whether it is net or gross energy consumption for which we are calculating the breakdown.

	Net	*Gross*
Population growth	39	30
Increased electricity use per capita	16	39
Increased per capita use of liquid transportation fuels (mainly autos and aircraft)	40	27
All other increased per capita uses	5	4

Table 2-4. Projected Regional Energy Consumption, by Sector and Energy Type, 1985

trillion Btu

	New York City	West-chester County	Nassau-Suffolk	Greater NYC Area	Rest of region	Total region
Residential						
Electricity	80.2	17.1	44.5	142.0	145.3	287.3
Gas	154.5	23.1	53.1	230.7	263.7	494.4
Other	234.5	68.5	177.0	480.0	294.4	774.4
Total	469.2	108.7	274.6	852.7	703.4	1,556.1
Commercial and public facilities						
Electricity	101.5	13.1	35.5	150.4	113.7	264.1
Gas	39.1	13.2	47.1	99.4	183.9	283.3
Other	275.2	38.0	92.7	405.9	372.9	778.8
Total	415.8	64.3	175.3	655.7	670.5	1,326.2
Industrial						
Electricity	39.2	4.3	12.8	56.5	78.6	135.1
Gas	12.4	2.1	5.8	20.2	129.1	149.3
Other	24.1	6.1	11.3	41.5	68.2	109.7
Total	75.7	12.5	29.9	118.2	275.9	394.1
Transportation						
Electricity	11.1	0.3	0.3	11.8	1.2	13.0
Other	263.9	75.6	181.3	498.4	874.9	2,074.6[a]
Total	275.0	75.9	181.6	510.2	876.1	2,087.6[a]
Total net energy consumption						
Electricity	232.0	34.8	93.2	360.6	338.8	699.5
Gas	206.0	38.3	106.0	350.3	576.7	927.0
Other	797.7	188.2	462.3	1,425.8	1,610.4	3,737.5[a]
Total	1,235.7	261.3	661.5	2,136.7	2,526.0	5,364.0[a]
Net total excluding electricty	1,003.7	226.5	568.3	1,776.1	2,187.2	4,664.5
Utility fuels[b]	n.a.	n.a.	n.a.	n.a.	n.a.	2,349.5
Gross energy consumption	n.a.	n.a.	n.a.	n.a.	n.a.	7,014.0

Source: From Table 18, *REC*, and reflecting the application of appropriate Btu conversion factors (see Appendix) to the physical data. See also text comments.

[a] These figures include what is necessarily a geographically unallocable component and is thus not shown separately. It is dominated by aviation, marine, and railroad fuels.

[b] See notes to Table 5-1 for definition.

Clearly, the combination of per capita electricity and transportation energy demand was the dynamic element during this period. It is interesting that little of the growth in overall energy consumption can be ascribed to per capita increases outside of electric power and transport (i.e., oil and gas use in homes, commercial enterprises, and industry) even though the prevailing level of such uses represents well over half of all energy consumption.

Looking to the future, it seems safe to say that NYR population growth will continue to slow down (see tables 2-1 and 2-2) in line with national trends. Even so, the forces making for increased energy consumption need not slow down proportionately, and could even accelerate. On the assumption that the birthrate will remain at its low 1970 level, and that immigration will continue its steady decline, the region's population is projected to increase by 14 percent between 1970 and 1985, averaging 0.9 percent yearly, compared with 1.2 percent during 1960–70. However, to the mid-1980s, fewer children and smaller households can nevertheless mean substantial growth in households (27 percent), employment (19 percent), and an accelerated growth of per capita income (a 65 percent rise projected). (Note, however, that, in contrast to total employment, NYR manufacturing employment is projected to decline absolutely.) In the absence of deliberate measures to cut demand, this future outlook translates into more of various energy-consuming services per capita, such as more air conditioning (only 40 percent of the region's households have it now), more housing floorspace (including second homes), more cars, and more air travel. It is in this light that the region's potential for future energy demand must be viewed.[3]

The projected trend in electric power growth deserves specific comment. Demand for electricity would rise from 86 million megawatt-hours in 1970 to 205 million in 1985, an estimate developed independently of the roughly comparable figure of 211 million projected by the electric utilities. Close to half of the increment in demand, reflecting an annual growth rate of 6 percent, would come from the residential sector. This projection is based on evolving past relationships among income, settlement density, and electric consumption for different counties of the region.[4] Historically, people have tended to purchase more electricity per dollar of income over time because utility bills, even at

---

[3] The historical data for these projections were "locked in" as of the early 1970s. Were we to consider trends to this year of publication (1975), the prospects for birthrates, population growth, and also for real income growth, would need to be dampened to some degree.

[4] Shown in *Regional Energy Consumption*, p. 9.

Table 2-5. **Average Annual Rates of Change in Regional Energy Consumption, by Sector and Energy Type, 1960–1970 and 1970–1985**

	New York City	West-chester County	Nassau-Suffolk	Greater NYC Area	Rest of region	Total region
Residential						
1960–70	−0.3	1.4	2.8	0.6	2.5	1.3
1970–85	0.8	3.8	6.9	2.6	2.5	2.5
Commercial/public facilities						
1960–70	2.6	1.8	3.0	2.6	3.0	2.8
1970–85	2.0	2.0	6.3	2.9	2.8	2.8
Industrial						
1960–70	0.5	5.4	3.5	1.5	2.6	2.2
1970–85	1.2	1.0	6.8	2.2	1.4	1.5
Transportation						
1960–70	2.4	3.4	7.1	4.1	3.4	4.6
1970–85	2.4	3.1	5.6	3.2	3.2	4.1
Total net energy consumption						
1960–70	1.0	2.3	4.1	2.0	2.9	2.8
1970–85	1.5	3.0	6.4	2.8	2.7	3.1
Electricity						
1960–70	5.2	7.4	9.5	6.2	8.3	7.2
1970–85	5.9	7.3	10.8	7.0	5.0	6.0
Gas						
1960–70	4.4	6.5	10.0	5.6	8.7	7.2
1970–85	2.1	3.0	9.9	3.4	4.1	4.0
Other						
1960–70	0.1	1.3	3.2	1.1	1.3	1.7
1970–85	0.6	2.4	5.2	1.9	1.8	2.5
Net total excluding electricity						
1960–70	0.7	2.0	3.8	1.6	2.5	2.4
1970–85	0.8	2.5	1.8	1.2	3.3	2.7
Utility fuels[a]						
1960–70	5.2	7.5	9.6	6.0	8.4	7.1
1970–85	n.a.	n.a.	n.a.	n.a.	n.a.	5.3
Gross energy consumption						
1960–70	1.9	2.9	4.7	2.6	3.8	3.4
1970–85	n.a.	n.a.	n.a.	n.a.	n.a.	3.5

Sources: Based on absolutes in tables 17 and 18, *REC* (showing data for 1960 and 1970); and Table 2-4 above. Note to Table 2-4 regarding unallocable component applies here as well.
[a] See notes to Table 5-1 for definition.

the high levels prevailing in the New York metropolitan area, have at least until recently tended to rise less than the consumer price index as a whole. Such a rise in electricity demand may be dampened in the future to reflect higher prices, but by assumption, as discussed earlier, this factor is left largely out of account in the present chapter.

Over one-third of the increment in electric power demand would come from commercial and public facilities, a fast growing sector, closely related to the economic growth of the region in general and growth in office floorspace (projected to increase 46 percent) in particular.[5] About one-sixth of the increment in demand for electricity would come from industry. This is based on a twenty-year trend in rising electricity use per worker, accompanied, however, by a decline in manufacturing employment. Lastly, the extension of the region's 264-mile rapid transit network by 29 miles in New York City and by 18 miles in New Jersey, and the air conditioning of more trains and of some subway stations would claim only about 1 percent of the added region-wide requirement for electricity between 1970 and 1985. (The subways' electric consumption now is only about twice the region's requirement for street lighting.)

On the whole, the electricity growth rate would decline from 7.2 percent annually in the past decade to 6 percent under this set of assumptions; but electricity's share of total net energy use would rise from 8.6 percent in 1970 to 13 percent in 1985. As a proportion of gross energy use, fuel inputs at power stations and in steam production would reach 34 percent, compared with 26 percent in 1970; and heat lost through conversion would represent 23 percent of gross consumption as against 19 percent in 1970.

Within the nonelectric component of total energy growth, demand for transportation fuels would increase from 1,130 trillion Btu in 1970 to 2,075 in 1985. About 53 percent of the increment would consist of highway fuels and 47 percent of aviation fuels, while the level of demand for railroad and marine use would stay at the present level. The highway fuel projection assumes auto registrations to rise from 6.86 million to 9.68 million based on recent relationships between autos per household, income and density, and the suburban pattern of population growth. By assumption, miles per gallon would continue their

---

[5] Specifically, potential electricity consumption was estimated county-by-county on the basis of its historic relationship to building density and employment. (Ibid.)

downward trend until 1977 and then level off, while annual miles per auto would stay constant at 9,400. Truck registration is assumed to remain constant in the core counties of the region and to continue growing at the 1960–70 rate outside, with annual gallons per truck remaining at the 1970 level. The aviation fuel projection assumes an increase in annual passengers from 37.4 million in 1970 to 93 million in 1985, a slowdown in the growth rate consistent with the "middle" projection in a recent RPA analysis;[6] the average trip length would increase by 30 percent, and the efficiency of airliners—reflecting engine performance and load factors of recent years—would be 12 passenger-miles per gallon. A factor is added for nonairliner traffic.

There remains nonelectric energy consumption in the household, commercial, and industrial sectors, the dominant source of demand being that for space-heating fuels in the household and commercial sectors. Growth of nonelectric power consumption in each of these is projected at about 2 percent yearly, while barely any growth at all is foreseen in the industrial sector. No doubt, nonelectric power growth for households and commercial structures would be dominated by liquid fuels, given evident supply constraints for natural gas.

In summary, the net energy growth rate would be 3.1 percent annually in the next fifteen years, not too different, given this set of projections, from the 2.8 percent rate in the past decade. (See Table 2-5.) Gross energy consumption would rise at 3.5 percent per year. However, only 24 percent of the potential absolute increase of 1,966 trillion Btu in net energy use would be attributable to population growth, compared with 39 percent during the 1960–70 decade (see foregoing text table); 19 percent would be due to higher electricity consumption per capita; and 17 percent to higher nonelectric consumption per capita in nonresidential facilities and households. As noted earlier, a number of forces underlying income and energy growth during the next decade—for example, the projected volumes of employment and building floorspace—are unaffected by slower population growth in the near term.

The following table summarizes the trend projections to 1985, by comparing the percent of net energy consumed in each sector of the NYR in 1970 with that projected for 1985.

[6] "The Region's Airports Revisited," *Regional Plan News*, No. 30, December 1973.

	Electric	Nonelectric	Total
		1970	
Residential	2.9%	28.6%	31.5%
Commercial/public facilities	3.3	22.3	25.6
Industrial	2.1	7.2	9.3
Transportation	0.3	33.3	33.6
Total	8.6	91.4	100.0
		1985	
Residential	5.4	23.7	29.0
Commercial/public facilities	4.9	19.8	24.7
Industrial	2.5	4.8	7.3
Transportation	0.2	38.7	38.9
Total	13.0	86.9	100.0

It can be seen that no really far-reaching shifts in principal sectors of use are indicated, if recent historic trends persist during the next ten years or so. The most marked change is in transportation demand, which would continue to raise its proportionate share of consumption; and this would also be the case with the share of the total delivered in the form of electricity.

A word should be added on the disparate energy growth that can reasonably be expected to prevail in subareas of the region. Table 2-5 shows how the approximately 3 percent growth rate in net energy consumption for the thirty-one county region as a whole fits into the wider range of growth trends projected within the region. New York City is seen to be at the low end of the growth spectrum, being projected at half the energy growth rate for the NYR in the aggregate. Suffolk and Nassau counties are well above the regional average. Other subareas, taken together, approximate the regionwide average, though, of course, examination of individual counties would disclose substantial variability in expected growth rates.[7]

Some Points Illustrated by the Figures

Clearly many factors quite apart from emergency restrictions on use (such as in early 1974) and the effect of price boosts could easily

[7] Individual county projections appear in *Regional Energy Consumption*.

combine to make actual demand fall short of these projections; for example, lagging income growth and less housing and office construction. Still, our exercise pinpoints the claim on energy resources inherent in the present structure of the region's economy, and can serve as a benchmark for evaluating strategies for change. From a conservation standpoint, several points seem pertinent.

1. The relatively rapid growth of electricity consumption in residential and commercial uses underscores the need to identify both the opportunities for, and constraints on, energy conservation measures in these sectors. Improving the efficiency of appliances, or impeding the shift to electricity for those uses where alternative sources of power are available, are among the paths that could be pursued.

2. The large share of energy consumption which, despite a very slow growth rate, will continue to be represented by space heating, poses a similar challenge.

3. The intensity of energy use varies inversely with urban density. To what extent should this phenomenon be reflected in long-term urban development policies?

We shall deal with a selected number of these key demand factors in subsequent portions of this report. The NYR's low per capita energy use (compared with the nation)—particularly in industry and transport —suggests that the payoff from conservation measures may well be more profitably pursued elsewhere in the country. But even at relatively low levels of per capita consumption there are such critical regional supply issues that conservation potentials and prospects should not be lightly dismissed. For example, reliance by the region's electric generating capacity on natural gas, though now shrinking, persisted even at a time when it was being denied to new customers, who were in a position to use it more efficiently. And we have mentioned the acute dependence of the region on imported residual oil. These supply problems prompt interest in the possibility of providing an alternative, environmentally acceptable fuel mix, with such sources as solid waste combustion, new uses of coal, nuclear systems coupled with pumped storage, and hydropower wheeled in from outside the region. Another possibility relates to the escalating amounts of waste heat from electricity generation, and ways of putting it to use, whether at central power stations or in decentralized total energy systems. Waste heat recovery has been much talked about but little practiced. These supply and technology questions, falling largely outside the scope of our own report, are being addressed in several other ongoing research projects.[8]

Summary   1886449

In this chapter we have sketched out the path along which regional consumption of fuels and power might unfold in the period to the mid-1980s in the absence of conscious energy conservation practices and policies. Demographic and economic indicators, "unconstrained" by a lack of energy, were projected for various parts of the region. The associated growth in fuel and power consumption was projected on the basis of historical relationships rather than on possible future departures from trend. For example, the impact of substantial increases in real energy prices was not taken into account at this stage of our effort. These calculations point to energy consumption growth slightly above that for the period since 1960. Electricity growth alone would, on these assumptions, decelerate but "only" down to an annual rate of increase of around 6 percent. No really far-reaching shifts in principal sectors of use are indicated, the most marked exception being some continued rise in the transportation sector's share of total energy consumption. The share of energy delivered in the form of electricity is projected to continue rising.

[8] For example: a project at the New York University Graduate School of Public Administration and a collaborative effort by the Brookhaven National Laboratory and State University of New York (Stony Brook). Both projects are supported by the National Science Foundation.

3.

A Review of
Energy-Conserving Practices

ONE can make a useful, if at times somewhat arbitrary, distinction between the nature of and opportunities for specific energy-conserving practices, on the one hand, and measures or policies likely to bring them about, on the other. Under the first category, we would, for example, include the fuel- or power-saving effect of increased attic insulation in homes; while, within the second category, we would group the legislated standards and/or market forces inducing the adoption of such practices. Thus, the first topic is oriented toward physical or engineering characteristics; the second, toward socioeconomic-political factors. In the same sense, the first category might well describe ideal conditions in which cost and other possible constraints are subordinated; while the second submits these conservation practices to the test of institutional and economic feasibility. In this chapter, we concern ourselves with a review of energy-conserving practices; in the following chapter, we turn to the matter of implementing the actions needed.

We begin with a schematic portrayal of energy conservation in its many different (and frequently misunderstood) meanings. Next, we look at the quantitative dimensions of some of the major energy-conserving practices, as these have been surveyed at the national level. Finally, we consider these in terms of their specific relevance to different parts of the New York Region.

Energy Conservation Schematized[1]

Proposals to dampen growth in energy consumption usually have the multiple objective of conserving finite resources, reducing imports, achieving savings in capital facilities, and minimizing environmental damage. For present purposes, let us assume two conditions as governing the pursuit of these goals: first, that for the time span for which it makes sense to look ahead, any suppression of energy-consumption growth will not be brought about by a deliberate slowdown of economic growth in general; and second, that a reduction in energy use is subjected to economic, and not merely physical or engineering, efficiency criteria. Thus, it seems questionable for a homeowner to bypass a conventional space conditioning setup in favor, say, of a heat-pump system featuring fewer Btu's of energy input per unit of effective energy output, if the shift involves a net increase in present or future costs. However, any such cost comparison should reflect or at least take into account environmental and other external costs traditionally not accounted for—in part because of quantification difficulties. With these conditions met, energy users can be accorded the privilege, for subjective reasons, to still act wholly contrary to what would appear to be their optimal economic stance.

Just as it seems proper to reject socially costly ways of reducing energy use, so it is necessary to point out that there are ways in which reduced energy consumption is essentially a by-product or, at best, only one element within a wider context of socially desirable policies. Thus, diminished reliance on large-horsepower private passenger cars reflecting low load-factor usage would very likely yield significant savings in energy consumption, but such a change in transportation arrangements is necessitated as much by overriding social needs as by efforts to cut down on fuel use.

To clarify the notion of energy demand dampening in terms of the multiple objective mentioned above, it may be useful to refer to the schematic outline of Table 3-1. The table distinguishes between the conservation of raw (or primary) energy resources, on the one hand, and energy as it is utilized by ultimate consumers, on the other. Table 3-1 also identifies the effect of various conservation practices on electric generating capacity, since much current controversy—particularly in

---

[1] The remarks in this section have been taken, with only minor modification, from my "Limiting the Demand for Energy: Possible? Probable?" *Environmental Affairs*, Vol. II, No. 4 (Spring 1973), pp. 717–731.

Table 3-1.　A Schematic Rundown of Energy-Conservation Practices

(x = energy saving; o = no energy saving)

Conservation practice	Conservation of:		
	Raw energy resources	End-use energy consumption	capacity Electric generating
1. Reduced population and income	x	x	x
2. Improved conversion efficiency (as in conventional electric generation)	x	o	o
3. Better load balancing	o[a]	o	x
4. Shift in energy end use towards form involving higher conversion efficiency (e.g., from electric to gas heating)	x	o	x
5. Shift in intermediate energy conversion towards a higher-efficiency technology (conceivably from conventional electricity generation to MHD or fuel cell)	x	o	o[b]
6. More efficient end-use energy utilization in satisfying given "need":[c]			
(a) Improved end-use technical efficiency, as in shift from incandescent to fluorescent lighting, lower horsepower automobiles, mass transit, or more efficient household motors	x	x	x[d]
(b) Reduced heat and light needs via improved building design and insulation	x	x	x[e]
(c) Eliminating waste (e.g., turning off unused lights, or raising summer thermostat when home is unoccupied)	x	x	x[e]
7. Shift towards less energy-intensive end-use activities:			
(a) Where purpose of a given activity can be achieved with greatly reduced fuel or power use (e.g., walking instead of riding, electronic communicating instead of traveling)	x	x	*f*
(b) Shift towards consumption of goods and services containing less embodied energy (e.g., more steel, less aluminum, natural instead of synthetic fibers)	x	x	x
(c) Tolerating increased discomfort (e.g., by waste heat utilization or by change in product–output specifications)[g]	x	x	x
8. Shift towards less energy-intensive, but still economic, production practices (e.g., by waste heat utilization, or by change in product–output specifications)[g]	x	x	x

Notes to Table 3-1

a Slight savings may accrue from not having to use inefficient peaking equipment.

b There may not be a saving in kilowattage, but perhaps one of site requirements.

c "Need" may have to be defined in physiological or normative terms.

d Except if the shift were to electrified mass transit, in which case electricity consumption (even if not energy consumption as a whole) would go up and so, therefore, would electric generating capacity.

e The extent of the saving in electric generating capacity depends on whether electric heating and cooling are involved.

f Effect is unclear.

g Less proliferation of models and increased durability could also produce raw energy savings.

the Greater New York City Area—concerns the issue of power plant siting requirements.

Some of the items cataloged in Table 3-1 (see line 1) reflect a point made earlier—that some opportunities for energy conservation are largely dependent on other social and economic trends. The desirability of those developments may be reinforced by the need to conserve fuels and power but have such widespread economic and social implications that it would appear inappropriate to suggest them solely as energy conservation measures.

Other lines in Table 3-1 illustrate the fact that conservation at one stage of energy flow need not signify savings at another point. Improving the efficiency of central electric generating stations, referred to in line 2, yields raw energy savings (say, of coal, oil, or gas) without any necessary diminution in final electricity utilization or, necessarily, of plant-site requirements. In this example, even in the absence of reduced consumer demand, resources are conserved; the environmental costs of producing them (say, oil spills or acid mine drainage) are mitigated, as are the thermal and air pollution emissions at power plants. But, unless generating efficiency improvements are such as to reduce the *number* of new plants needed, the esthetic disruption arising from the presence of a power plant or transmission line remains. Line 5 is similar to line 2 insofar as it illustrates the effects of enhanced central station conversion efficiency.

Line 3 reminds us that offpeak energy use—turning on the electric clothes dryer at midday rather than during the hours of intensive power demand—may reduce the volume of required generating and transmission capacity. It does not, however, reduce the quantity of raw energy needed for combustion[2] or the amount of electricity consumed.

Line 4 reflects the fact that some forms of energy utilized by the consumer are judged to represent a lesser drain on primary energy resources than other forms. An example commonly cited suggests that by delivering heat to households from gas furnaces rather than through electric resistance heating units, primary energy savings occur because the low-efficiency electric plant conversion stage is thereby bypassed. The fact that the conservation benefits of switching away from electric heat seem frequently to be overstated (because of failure to acknowledge heat loss and poor maintenance of alternative heating systems, as discussed below), and the arguments for such a shift oversimplified, does not alter the basic point of this example.

The components of lines 6 and 7 are those cited most frequently in discussions of dampening energy demand. Those subsumed under line 6 are examples of how given types of consumer needs might be accommodated with less energy at both the primary and final consumption stages. The line 7 examples presuppose a shift in expenditure patterns or life styles. The desirability of such changes is, of course, highly subjective.[3]

Line 8 raises the possibility of energy conservation opportunities in production activity. These opportunities could span a range of things—eliminating unnecessary lighting in warehouses, stopping steam leaks, instituting improved combustion control. In the past, such industrial energy-saving potentials, and the associated cost implications, have been rather neglected, but they are now beginning to receive serious attention at a number of places (e.g., as part of a research effort under way at the National Bureau of Standards in Washington).[4]

---

[2] Except insofar as peaking power frequently involves the deployment of least efficient plant, which can be half as efficient as regular base-load facilities.

[3] The nature of the spatial distribution of the population is one aspect of life-styles having perhaps significant consequences for energy use and energy conservation possibilities. More scattered development means more transmission line losses, and more energy consumption for transportation, water and sewer pumping, and waste collection.

[4] Some persons perceive great virtues in switching from what they label lower-power-productivity industries (such as aluminum) to high-power-productivity in-

Very often, the problems associated with energy consumption are, as we stress in this report, of a particularly regional or localized nature —for example, where to locate additional generating capacity for New York City. Clearly, regional conservation practices may not signify national energy savings and vice versa. For example, a regional limit to expanded electricity capacity and generation which induces industrial relocation simply shifts the burden of adjustment elsewhere; conversely, national energy savings achieved, say, by switching away from energy-intensive production (such as plastics or aluminum) obviously may have little or no impact within a given region.

Some might want to broaden the conception of energy conservation reflected in Table 3-1 still further by alluding to opportunities at the energy raw-materials production end. Significant advances in petroleum recovery techniques, for example, can help stretch out energy supplies that might otherwise have been lost to society and, conceivably, make less pressing the need that would otherwise exist for energy-saving habits in household, industrial, or transport usage. The development of novel types of primary energy, among which solar power is a definite possibility, falls into a similar category. But such an extension of the term blurs rather than illuminates the concept.

Having illustrated—in this brief cataloging—the full range of energy-conserving practices, let us return to those aspects addressed in the present study. In this report, we concern ourselves largely with energy-conservation potentials illustrated by lines 3, 4, 6, and 7 of Table 3-1— that is, with questions bearing on the efficacy of greater energy-conserving practices at only the *end-use* consumption stage. We are

---

dustries (such as steel). See B. Commoner, paper delivered to the American Association for the Advancement of Science, Philadelphia, Pennsylvania, December, 1971. While such shifts may curtail energy use, the question of whether such a shift is resource saving and environmentally benign from an *overall* perspective is far more difficult to answer. For example, steel production requires coal and iron ore mining and haulage, and blast furnace and coke oven operation; these include high, partially external, environmental costs. Switching away from low-power-productivity industries (often the most capital-intensive, the most mechanized, and featuring the highest labor productivity) to less capital-intensive, less mechanized, more labor-intensive industries is, to say the least, a mildly bizarre twist to the usual conception of economic and social development.

interested in exploring ways in which ultimate users of energy in the household, commercial/public facility, industrial, and transportation sectors of the economy[5] can profitably (and relatively innocuously) reduce their energy consumption, whether by improving the efficiency of a given type of energy use or by shifting toward a more efficient use pattern. As noted at the outset of the chapter, we ignore, as something outside our particular perspective, the possibility that suppression of energy-consumption growth might be brought about by a deliberate slowdown of economic growth in general; in other words, we want to deal with opportunities for less energy consumption at given levels of "welfare," however that notoriously elusive term is defined. We also bypass, for the most part, consideration of major technological advances as a means of economizing on energy (see lines 2 and 5 of Table 3-1); for example, while the prospects for improving the load factor for private passenger cars definitely concerns us, energy implications of new types of automotive power plants are of no direct concern here.

*Potential Conservation in the Major
Consumer Sectors: A National Overview*[6]

Opportunities for conservation in the use of energy have been reported in a series of studies released during the past several years. To the extent permitted by the nature of the underlying data—and these are frequently quite deficient—these reports evaluate energy savings possi-

[5] Electric utilities (where conservation potentials reside largely in the field of advanced conversion technology) are viewed here as an intermediate supplier, rather than an end user, of energy. Some electricity utility questions are, however, touched upon briefly in the context of residential and commercial total energy systems.

[6] Much of the following section has been adapted from Joel Darmstadter and Eric Hirst, "Energy-Conservation Research Needs," in Hans Landsberg, and others, *Energy and the Social Sciences—An Examination of Research Needs*, RFF Working Paper EN-3 (Resources for the Future, 1974). This study was funded by the National Science Foundation. For the most part, the conservation literature cited here represents publications available during 1972–73. It should be acknowledged that voluminous additions to this literature have appeared more recently—especially under the auspices of such federal government organizations as the Federal Energy Administration, the Department of Transportation, the Department of Housing and Urban Development, the Federal Power Commission, the Environmental Protection Agency, the Council on Environmental Quality, the National Bureau of Standards, and numerous congressional committees.

Table 3-2. Studies Dealing with Strategies for Increasing Efficiency of Energy Use

Strategy	References (see below)
Transportation	
Shift modal mix to energy-efficient modes	1,2,3,4
Increase load factors, especially for urban travel	1,2,4
Improve vehicle energy efficiency for autos, trucks, airplanes	1,5,6,4
Residential and Commercial	
Increase building insulation to reduce heating and cooling needs	7,3,8
Use heat pumps or fossil-fired heating systems rather than resistance heat	3,9,8
Use energy-efficient air conditioners	3,8
Use total energy systems to provide both heat and electricity	10,11
Improve energy performance of appliances	8
Pay closer attention to lighting levels	9
Industry	
Increase recycle of energy-intensive materials	12,13
Improve energy efficiency of industrial processes	14,15
Use waste heat from power plants for direct heat and process steam	16,17
Use solid wastes as fuel	17

References
1. U.S. Office of Emergency Preparedness, *The Potential for Energy Conservation—A Staff Study* (Washington: October 1972).
2. E. Hirst, *Energy Intensiveness of Passenger and Freight Transport Modes: 1950–1970.* Oak Ridge National Laboratory, Report ORNL-NSF-EP-44 (April 1973).
3. E. Hirst and J. C. Moyers, "Efficiency of Energy Use in the United States," *Science,* vol. 179, no. 4080 (March 30, 1973).
4. E. Hirst, "Transportation Energy Use and Conservation Potential," presented at Conference on Energy: Demand, Conservation and Institutional Problems, Massachusetts Institute of Technology (Cambridge, Mass.: February 12–14, 1973).
5. W. S. Anderson, *R&D for Fuel Economy in Automotive Propulsion,* U.S. Army Tank-Automotive Command (June 19, 1972).
6. U.S. Environmental Protection Agency, Office of Air and Water Programs, *Fuel Economy and Emission Control* (Washington: November 1972).
7. J. C. Moyers, *The Value of Thermal Insulation in Residential Construction: Economics and the Conservation of Energy,* Oak Ridge National Laboratory, Report ORNL-NSF-EP-9 (December 1971).
8. D. G. Harvey and others, "Minimization of Residential Energy Consumption," in *Conference Proceedings: 7th Intersociety Energy Conversion Engineering Conference* (Washington: American Chemical Society, 1972), pp. 1247–55.
9. R. G. Stein, "A Matter of Design," *Environment,* vol. 14, no. 8 (October 1972).
10. M. G. Gamze, "A Critical Look at Total Energy Systems and Equipment," in *Conference Proceedings: 7th Intersociety Energy Conversion Engineering Conference* (Washington: American Chemical Society, 1972), pp. 1263–68.
11. P. Achenbach and others, "Operation Breakthrough—Total Energy," Effective Energy Utilization Symposium (Drexel University, Philadelphia: June 8–9, 1972).
12. J. Bravard and others, *Energy Expenditures Associated with the Production and Recycle of Metals,* Oak Ridge National Laboratory, Report ORNL-NSF-EP-24 (November 1972).
13. R. G. Hunt and W. E. Franklin, "Environmental Effects of Recycling Paper," presented at 73rd National Meeting of American Institute of Chemical Engineers (Minneapolis: August 27–30, 1972).
14. B. Commoner, "Power Consumption and Human Welfare," presented at the Annual Meeting of the American Association for the Advancement of Science (Philadelphia: December 29, 1971).
15. C. A. Berg, *Energy Conservation Through Effective Utilization,* National Bureau of Standards NBSIR-73-102 (Washington: February 1973).
16. A. J. Miller and others, *Use of Steam-Electric Power Plants to Provide Thermal Energy to Urban Areas,* Oak Ridge National Laboratory, Report ORNL-HUD-14 (January 1971).
17. U.S. Environmental Protection Agency, *Energy Recovery from Waste* (Washington: 1972).

bilities in considerable detail. Table 3-2 lists major studies which have appeared. Numerous additional projects in the conservation area have more recently been, or are soon to be, published—for example, studies commissioned by the recently completed Energy Policy Project of the Ford Foundation.

Residential Sector. Space heating is the largest energy user in homes, accounting for close to two-thirds of the nation's residential energy "budget." It is also of key importance in the New York Region's use of energy (see Table 1-5 above and the accompanying text). Water heating, with 14 percent of the total, is the second largest. Cooking and clothes drying account for probably an additional 6 percent or so of residential energy consumption. Thus, the use of energy for direct heat represents about four-fifths of the energy consumed in homes. The remainder is used for refrigeration, air conditioning, lighting, and operation of small appliances.

Because space heating is such a large energy user, much research has concentrated on ways to reduce space heating energy requirements through increased building insulation, improvements in design of heating equipment, and building design changes. One report examined the energy and dollar savings due to the use of ample thermal insulation and storm windows for single-family homes.[7] The results show that insulating the home to minimize total life-costs of heating plus insulation would reduce space heating energy needs by at least 40 percent relative to the 1965 Federal Housing Administration standards, less so compared with more recent (upgraded) standards. As fuel prices increase, additional levels of insulation will be economically justified; this would further decrease space heating energy use.

Adding insulation to multifamily homes is also likely to save energy. Work under way at Hittman Associates, the National Bureau of Standards, and Princeton University[8] is designed to quantify these savings.

[7] J. C. Moyers, *The Value of Thermal Insulation in Residential Construction: Economics and the Conservation of Energy*, Report ORNL-NSF-EP-9 (Oak Ridge: Oak Ridge National Laboratory, December 1971).

[8] See Hittman Associates, *Residential Energy Consumption, Phase I Report*, HUD-HAI-1 (March 1972); also, *Residential Energy Consumption, Multi-Family Housing Data Acquisition*, HUD-HAI-3 (Washington, October 1972); D. G. Harvey and others, "Minimization of Residential Energy Consumption," presented at 7th Intersociety Energy Conversion Engineering Conference, San Diego, September 25–29, 1972 (Washington: American Chemical Society, paper no. 729187); C. A. Berg, *Energy Conservation Through Effective Utilization*, National Bureau of

Some of this research is also related to the energy impacts of design parameters, such as building geometry and orientation.

Energy consumption for space heating depends not only on building design but also on the type of heating equipment used. There is considerable controversy over the relative environmental, economic, and energy conservation merits of gas- or oil-fired heating systems, electric resistance heaters, and electric heat pumps. Heat pumps are about twice as energy-efficient as electric resistance heaters on a national average basis,[9] but there is considerable regional variation in the performance of heat pumps as well as in their competitive attractiveness. Also, resistance heaters can be adjusted room-by-room. Gas or oil heaters are judged more energy-efficient than electric resistance heaters, though they are themselves subject to greater or lesser energy losses. Thus questions concerning the relative energy standing of different heating systems have not yet been adequately resolved.

Although water heating accounts for a sizable portion of the residential energy budget, very little work has been done on methods for improving its energy efficiency. As with space, water can be heated either electrically or by direct fossil-fuel combustion. From the standpoint of raw-energy input, gas water heaters are about twice as energy-efficient as electric heaters,[10] but more work needs to be done to support this estimation. One report found that three inches of additional insulation on electric water heaters was economically justified at 1972 prevailing electricity prices in the TVA region—traditionally a low-cost electricity region.[11] This increased insulation provides an estimated energy saving of somewhat under 10 percent. Comparable or perhaps larger savings would accrue from more effective insulation of gas water

---

Standards NBSIR-73-102 (Washington, February 1973); "Environmental Testing of Full-Size Buildings," *U.S. Department of Commerce News* (Washington, February 1973); R. A. Groot and R. H. Socolow, *Energy Utilization in a Residential Community*, Princeton University Center for Environmental Studies, Working Paper W-7 (February 20, 1973).

[9] J. C. Moyers, "Analysis of Energy Uses," in *Electrical Energy and Its Environmental Impact—Progress Report, December 31, 1972*, Report ORNL-NSF-EP-40 (Oak Ridge National Laboratory, March 1973).

[10] Stanford Research Institute, *Patterns of Energy Consumption in the United States*, Report to Office of Science and Technology, Executive Office of the President (Washington, 1972).

[11] R. S. Quinn, Jr., "The Effect of Increased Capital Expenditures as a Method of Reducing Electricity Demand for Hot Water Generation in a New Home," M. A. Thesis, University of Tennessee, Knoxville, March 1973.

heaters. To our knowledge, that question has not been dealt with. Quinn also examined the economics of insulating hot water pipes and of pre-heating incoming cold water by a heat exchanger installed in the attic. However, neither of these schemes has seemed economically justifiable —at least not at electricity prices prevailing in the early 1970s.

Energy savings in the design of several appliances are possible. For ranges, a shift to gas units, increased oven insulation, and improved oven-door design could save energy.[12] Insufficient data are available to detail the potential energy savings through increased refrigerator insulation and condenser/compressor design. Apparently, some refrigerators have such thin walls that small heaters are installed to prevent condensation on the outside walls of these models; this results in a double energy penalty.[13]

Air conditioning is currently still a relatively minor factor in total energy use in homes, though it figures more importantly in residential electricity use and more importantly still as a contributor to peak electric power loads in the New York Region and elsewhere. It is also the fastest growing type of residential energy use, with an average annual growth rate in energy use of over 10 percent.[14] The energy efficiencies (i.e., the ratio of Btu per hour of cooling divided by watts of power consumed) of all window units sold in 1972 varied by more than a factor of two—from 4.7 to 12.2 Btu/Watt-hour,[15] with an average figure of around 6.0. There seemed to be little correlation between energy efficiency and retail price. To our knowledge, a similar examination of energy efficiency for central air conditioning units has not yet been con-

[12] See Hittman Associates, *Residential Energy Consumption*, and D. G. Harvey and others, "Minimization of Residential Energy Consumption."

[13] It is worth noting, in this respect, that the Philco-Ford Corporation has announced a new line of refrigerators described as using 30 to 50 percent less electricity than existing models. The new units embody improved insulation, permitting use of smaller motors and compressors and eliminating the use of the heaters referred to in the text. At 3 cents/kwh for a 615-Watt, 14-cubic foot unit consuming an estimated 1,829 kwh/year, there is said to be a reduction in annual electric power costs from $73 to $51 down to $35, and undiscounted lifetime savings ranging from $267 to $607. No price increase was said to be contemplated compared with earlier models. (*New York Times*, March 13, 1973, pp. 49, 63.)

A comprehensive project looking at all energy-related aspects of refrigeration has been under way at the MIT Center for Policy Alternatives, under the direction of J. H. Hollomon.

[14] Stanford Research Institute, *Patterns of Energy Consumption*.

[15] J. C. Moyers, "Analysis of Energy Uses."

ducted. Insulation, in addition to reducing heating requirements, also cuts air conditioning requirements The optimum level of insulation discussed above[16] would reduce air conditioning energy consumption by about 15 percent relative to the 1965 Federal Housing Administration standards, and somewhat less relative to more recent standards.

Gas pilot lights offer additional energy-saving opportunities. Their continuous consumption in a typical house has been estimated at 4 percent of the total gas needs of a gas-heated residence.[17] Particularly for ranges and clothes dryers (which are used only a small fraction of the time), pilot light energy losses are quite significant in terms of the energy required by these appliances. Use of electronic igniters would save energy but increase the initial cost of these appliances. They are probably not feasible for existing units. For space and water heaters, the pilot light energy losses are less severe because some of their heat is utilized.

In addition to examining individual energy users, the Hittman Associates and Harvey also discussed a number of ways to improve the overall energy efficiency of homes by using waste heat from gas flues, waste hot water, clothes dryer discharge, and refrigerator condenser waste heat. Their results, though quite tentative, suggest that large energy savings are possible, but only with significant changes in the design of homes.

The foregoing review shows that considerable work has been done or is under way dealing with energy end use and energy conservation possibilities within the home. Additional "hardware" research is needed to define accurately overall energy efficiencies for gas versus electric appliances and space heating equipment, including examination of the effects of maintenance and deterioration; and to define the variations in energy efficiency among different models of the same appliances (as discussed above for air conditioners). Much of this work can probably be conducted within the industries producing these appliances.

Commercial Sector and Public Facilities. This grouping can be broadly defined as that portion of the economy, which, for the most part, delivers services (other than transportation) rather than goods.

---

[16] J. C. Moyers, *The Value of Thermal Insulation.*
[17] M. R. Seidel, S. E. Plotkin, and R. O. Reck, *Energy Conservation Strategies,* U.S. Environmental Protection Agency (Washington, July 1973), p. 62.

This sector includes: construction; communications; wholesale and retail trade; finance, insurance, and real estate; personal services (hotel, repair, recreation, legal, educational); and government.[18] However, agricultural energy use is also included in the commercial sector. Because of its diversity, it is almost impossible to sensibly define commercial energy-use patterns. Nonetheless, the commercial sector to a large extent is dominated by energy uses that are roughly comparable to those in the residential sector: the predominant portion of the energy is used for space heating of buildings, and much smaller fractions are used for water heating, air conditioning, cooking, and refrigeration.[19] (See Table 1-5.) However, lighting is proportionately a much more important component of total energy use in the commercial sector than in homes. Energy-use data for agriculture are scanty; although accounting for little "commercial" energy consumption in percentage terms, energy can be of considerable importance in particular agricultural processes.

To the extent that commercial energy-use patterns are similar to those in the residential sector, the potential energy-conservation measures reviewed above may be applicable here. The National Bureau of Standards has been conducting research[20] which should quantify various potential energy savings in large office buildings. Several architects are engaged in similar studies.[21] Preliminary results show possible energy savings in large buildings through closer attention to lighting levels, reduced ventilation, and use of devices to extract heat from building exhaust air.

Work under way for the U.S. Department of Housing and Urban Development is evaluating the potential of total energy systems.[22] Such

[18] U.S. Bureau of the Budget, *Standard Industrial Classification Manual*, Washington, 1967. (It should be recalled, from notes to Table 1-1, that national energy data issued by the U.S. Bureau of Mines include, within the commercial, industrial, and transport sectors, nonfuel uses of energy commodities—e.g., petrochemical feedstocks or asphalt and road oils used in road construction. We have, for the most part, excluded these from our statistics in order to provide comparability between national and regional data since we could not develop regional estimates on such products).

[19] Stanford Research Institute, *Patterns of Energy Consumption*.

[20] C. A. Berg, *Energy Conservation* and "Environmental Testing."

[21] R. G. Stein, "A Matter of Design," *Environment*, vol. 14, no. 8 (October 1972); F. S. Dubin, "Energy Conservation Through Building Design and a Wiser Use of Electricity," Annual Conference of the American Public Power Association (San Francisco, June 26, 1972).

[22] See M. Gamze, "A Critical Look at Total Energy Systems and Equipment,"

systems generate electricity for apartment houses, shopping centers and the like, and use what would otherwise be waste heat for space and water heating and absorption air conditioning.

Industrial Sector. This sector comprises the nonagricultural goods producing part of the economy, i.e., manufacturing and mining. The preeminent share of energy usage in industry involves the application of direct heat and process steam, although, within the present study, that share (see Table 1-5) is somewhat higher than it would be were we also to take into account the utilization of energy raw materials in applications other than fuels and power. (See notes to Table 1-1.) About half of industrial energy consumption is concentrated in six manufacturing activities: primary metals; chemicals; stone, clay, and glass; paper; food; and petroleum refining (including coal products).

There are perhaps three broad routes along which to proceed in the search for ways to economize in industrial energy use. The first is to examine methods for increasing energy efficiency of existing industrial processes; for example, through improved furnace insulation or maintenance. A second approach involves examination of alternative processes for producing particular products and the energy consequences of these alternatives. The use of waste heat from central station electric power plants for industrial process steam is such a possibility, as is the recapture of heat from industrial boilers and converters. Finally, the energy implications of product substitution and of producing goods for long lifetimes and low maintenance need to be examined.

The foregoing are acknowledged to be little more than generalizations. There is a need for research to disclose significant conservation possibilities in *specific* industries and processes. For example, a new Alcoa process announced in early 1973 points to the possibility of a 30 percent reduction in electricity requirements, implying a big efficiency improvement in aluminum making.[23]

But even short of technological breakthroughs—which is what is implied in the last example—substantial energy savings are believed

---

in *Conference Proceedings: 7th Intersociety Energy Conversion Engineering Conference* (Washington: American Chemical Society), pp. 12, 63, 68. Achenbach and others, "Operation Breakthrough—Total Energy," Effective Energy Utilization Symposium (Drexel University, Philadelphia, June 8–9, 1972).

[23] *Chemical Engineering Journal,* January 15, 1973, p. 12; and February 26, 1973, p. 11.

realizable with existing technology. "Since losses of energy do not normally show up in the same manner as off-quality production, inefficient labor, and equipment breakdown, huge losses are often tolerated simply because no one knows they exist."[24] This suggests the desirability of introducing energy "audit" as an explicit and standard element of cost accounting in industrial operations.

Transportation Sector. The automobile uses a larger share of transport energy than all other passenger and freight modes combined, around 55 percent in recent years. (We are not considering here energy embodied in, or needed to produce, transportation equipment.[25]) Trucks consume an additional 20 percent of the national transportation energy total. Commercial aviation (especially passenger service) is the fastest growing transportation energy user, having increased its share of the transport energy total from under 3 to approximately 7 percent in a decade. The fraction of transportation energy devoted to passengers has climbed steadily over the past ten to twenty years, reaching about 60 percent during the early 1970s. The remainder is used for freight, military transport, and miscellaneous purposes (Table 1-5).[26] The composition of the NYR's energy budget in transportation follows essentially the same rank-ordering as in the national pattern, except that forms of public transport (bus, train, subway) are proportionately more important in the local situation.[27]

Basically, there are three ways to increase energy efficiency in transportation: shift traffic to energy-efficient modes, increase transport system load factors, and improve energy efficiency for individual modes. Several papers describe potential energy savings due to shifts in

[24] G. F. Moore, "Energy Management in the Industrial Community," remarks at National Energy Forum, sponsored by U.S. National Committee of the World Energy Conference, Washington, March 18, 1973.

[25] For such calculations, see Eric Hirst, *Direct and Indirect Energy Requirements for Automobiles*, Report ORNL-NSF-EP-64 (Oak Ridge: Oak Ridge National Laboratory, February 1974).

[26] See Eric Hirst, *Energy Intensiveness of Passenger and Freight Transport Modes: 1950–1970*, Report ORNL-NSF-EP-44 (Oak Ridge: Oak Ridge National Laboratory, April 1973); and W. E. Mooz, *The Effect of Fuel Price Increases on Energy Intensiveness of Freight Transport*, Report R-804-NSF (Santa Monica: The Rand Corporation, December 1971).

[27] *Regional Energy Consumption*, Second Interim Report of a Joint Study by Regional Plan Association, Inc. (RPA) and Resources for the Future, Inc. (RPA, New York, N.Y., 1974).

modal mix for both passengers (from airplanes and autos to trains, buses, and mass transit) and freight (from airplanes and trucks to trains).[28] In general, the modal shifts hypothesized in these papers are discussed without reference to time-phasing (i.e., no indication is given of the time required to implement such shifts). More important, the capital requirements and the mechanisms required to induce such shifts (which would actually be counter to recent trends) are not discussed. This should not be interpreted as a criticism of these papers; implementation of energy-conserving transportation systems is likely to be quite complicated.

Little work has been done to assess the possibilities of, and energy savings due to, improved transport system load factors. Improved load factors might be particularly important for urban passenger systems in which, on the average, only one seat in four is occupied.

Considerable work is under way to evaluate new technologies and applications of existing technologies which would reduce fuel consumption for transportation vehicles.[29] For example, depending on load factors, small cars and jumbo jets consume less fuel per seat-mile than do large cars and conventional jets. This does not argue for small cars and jumbo jets as the transportation underpinnings of society. It only points to the need for research which would attempt to evaluate the incentives and obstacles to use of such energy-efficiency alternatives or of possible offsetting costs.

*Potential Conservation in Selected Major Use Categories:
The Regional Picture*

In turning to a survey of significant energy conservation opportunities in New York City and its multicounty surroundings, we have to rely on rather fragmentary evidence, both with respect to the specific

---

[28] Three useful references are: E. Hirst, "Transportation Energy Use and Conservation Potential," *Bulletin of Atomic Scientists* (November 1973), pp. 36–42; A. C. Malliaris and R. L. Strombotne, "Demand for Energy by the Transportation Sector and Opportunities for Energy Conservation," presented at Conference on Energy: Demand, Conservation, and Institutional Problems, MIT, Cambridge, Mass., February 12–14, 1973; and R. A. Rice, "Energy Efficiencies of the Transport Systems," Society of Automotive Engineers, Annual Congress, Paper No. 730218, Detroit, January 8–12, 1973.

[29] U.S. Environmental Protection Agency, Office of Air and Water Programs, *Fuel Economy and Emission Control* (Washington, November, 1972).

components of total energy consumption for which data are available and to the limited parts of the thirty-one-county tristate region to which such information applies. Most of the existing documentation refers to New York State (usually in the form of statewide aggregates) so that generalization of these data to the thirty-one-county grouping or to the narrower New York City area is, at best, a matter of rough inference. However, since our purpose is not so much one of precise estimation as of "order-of-magnitude" conjecture, this limitation need not be severely cramping. Both because of their dominance within the NYR's total energy consumption (see Chapter 1) and their susceptibility to more efficient usage, space heating (considered here along with cooling, because of many overlapping aspects) and transportation are singled out for emphasis.

Space Heating and Cooling. For shorthand, we label this the "space-conditioning" function. Improved thermal insulation of buildings would be a key move towards enhanced efficiency in this energy use. Numerous studies have disclosed beneficial economic trade-offs resulting from a greater initial investment in insulation than has traditionally been undertaken in new construction. That is, the incremental capital costs associated with improved insulation are more than recaptured by reduced fuel bills early in the life of the structure; or, putting it still another way, "annualized"[30] total energy costs are minimized (energy use is economically optimized) by conforming to ideal insulation practices. Nonetheless, these economies are not yet widely acted upon. As a recent report surveying the situation on New York State observes: "The amount of thermal insulation installed in a new home heated with gas or oil is generally about 50 percent of that which would be prescribed based on a total annual cost comparison. Builders tend to minimize the initial cost of insulation at the expense of the non-suspecting home purchaser."[31]

Generally speaking, only electric resistance space heating systems are associated with anything approaching optimal insulation practices. Thus, the electrical industry has, since the 1950s, subscribed to the so-called

[30] An annualized charge refers to the amortized yearly cost based on the amount of the initial investment, the life of the equipment, and the interest forfeited by the sunk capital.

[31] *Report of the Ad Hoc Committee on Appliance and Apparatus Efficiency* to the Interdepartmental Fuel and Energy Committee of the State of New York (Albany, New York, June 25, 1973), p. 260.

all weather comfort standard for electrically space-conditioned homes —a standard judged a guide to good thermal performance values and vapor barrier and ventilation practices. In 1970, somewhat under 2 percent of total occupied housing units in New York State had electric heating. This constituted under 7 percent of total residential electrical energy usage or about 2 percent of total electrical energy usage in New York State in that year.[32] In 1971, the Federal Housing Administration mandated maximum heat losses for new one- and two-family homes using *any type of fuel* that were the same as had previously governed only electrically heated homes. However, since it applies only to federally assisted housing, which is only about 6 percent of the total for new construction in New York State, the revised FHA insulation standard can contribute only modestly to the incorporation of sound installation in new housing—particularly in multifamily housing;[33] it has no impact on the existing housing market. But this leads quickly to the question of the means by which energy saving practices, such as improved insulation, are to be implemented, and the point in time at which they can be assumed to have a measurable impact on overall energy use. We shall defer this topic to the next chapter. Suffice it to say that, while it is important to demonstrate that space-conditioning energy use in the New York Region might, with proper insulation, be reduced by, say, 25 percent (implying perhaps a 10 percent reduction in total net energy use), this does represent a somewhat idealized calculation, lacking in short- and perhaps intermediate-term practical significance.

Nevertheless, the longer-term benefits of improved insulation to the nation and NYR, and the earlier payoffs to the individual homeowner, must be underscored. In testimony before the New York State Public Service Commission regarding energy conservation issues in the Con Edison area, John Moyers (an engineer from Oak Ridge National Laboratory) summarized the net yearly dollar savings—i.e, the difference between annualized insulation costs and fuel savings—resulting from economically optimum insulation practices,[34] as follows:

[32] Ibid., pp. 54–55.

[33] Most new single-family homes probably will meet FHA requirements since builders wish to allow buyers to obtain FHA financing. Multifamily housing, with different financial arrangements, is probably built on a least-first-cost basis.

[34] State of New York, Public Service Commission, *Report on Energy Conservation in Space Conditioning*, Case 26292, Report by William K. Jones, Commissioner (January 31, 1974), pp. 180–83. See this source for additional specific

	New home with gas heat	New home with gas heat and electric air conditioning
Net annual savings	$54.40	$61.70
Associated energy savings:		
Gas, thousand Btu (percent of gas for heating)	39,700 (29%)	39,700 (29%)
Electricity, kwh (percent of electricity for air conditioning)	—	150 (14%)

The data apply to a new, medium-sized, single-story home in the New York City area, as of the early 1970s. A recalculation of Moyers' study using up-to-date (1974) data concluded that the rising cost of insulation and labor has approximately kept pace with the rise in fuel costs, so that the cost-effectiveness of insulation remains about what it was at the time of Moyers' initial investigation.[35] Energy costs assumed in Moyers' original calculations were 2.2¢ per kwh and $1.50 per million Btu of gas. The implied pay-back time of the initial investment appears to be not much over five years. This suggests a high rate of return, keeping in mind that insulation, once installed, can be expected to retain its effectiveness for the life of the structure.

Improved operation and maintenance of furnaces and air conditioners may also yield energy savings. Opportunities here could potentially be diffused much more quickly than in the insulation case (though quantitatively less significantly) both because they need not involve complicated and costly retrofitting and because the life-span of the equipment, being considerably less than that of homes, signifies a rapid enough turnover to produce noticeable economies in energy use in a matter of years rather than decades.

Residential gas heating appliances normally provide for minimum thermal efficiencies of around 70 percent (i.e., a 30 percent heat loss

---

assumptions underlying Moyers' analysis. In his testimony, Moyers also indicated that such retrofitting as would seem feasible in existing homes would yield economically worthwhile savings.

[35] Robert T. Taussig and others, "Energy Conservation in High Density Areas: A Study of the Energy Crisis in New York City," Columbia University School of Engineering and Applied Science, September 20, 1974, p. 53 (unpublished).

between input and output). However, utilization efficiency is frequently as low as 40 percent due to incorrect appliance selection or imperfect operation and maintenance. Much the same situation characterizes oil furnaces. Indeed, much of the controversy regarding relative energy efficiencies of gas, oil, and electric furnaces—already alluded to—turns on operation and maintenance assumptions.

As noted, the energy-use implications of the type of heating system that is chosen for a new residential dwelling unit has generated intensive debate over the virtues of electric space heating versus gas- or oil-fired furnaces. The former, though highly efficient at point of use, requires substantially more raw energy than is the case with direct fossil-fuel heating, which bypasses the electric conversion stage. When gas- or oil-fired furnaces are operated optimally, their 70 percent efficiency contracts with the best central generating stations of 35–40 percent efficiency. Theoretically, therefore, direct combustion may imply as much as a 50 percent fuel input saving. It is this fact which, at the conclusion of recent hearings, prompted a New York State Public Service Commissioner to recommend prohibition of electric resistance heating in new construction, except under limited circumstances.[36]

The controversy seems unlikely to abate quickly, however. Proponents of electric heating (including some who are not bound by an advocacy posture) make the points that (a) alternative systems often are, in fact, operated at poor efficiency, (b) a number of inevitable heat losses occur, (c) present and prospective oil and gas shortages may force greater-than-expected reliance on coal- or nuclear-based electricity (see, for example, the demand management options of the *Project Independence Report*),[37] and (d) advanced pollution controls at a modern electric generating station may result in less environmental damage than highly dispersed individual home heating systems. But while the banning of new electric heating installations would not now have a decisive impact on total energy in New York State, these installations have been growing rapidly in the state—over 10 percent annually in both 1971 and 1972—and their prohibition could in time be rather important.

As a technological alternative to electric heating, the most attractive, though not novel alternative, is the heat pump, which, if claimed effi-

[36] Ibid., p. 284.
[37] Issued by the Federal Energy Administration, Washington, November 1974.

ciencies are to be believed, could reduce average electric space heating energy requirements by around 50 percent. Heat pumps reverse the operation of air conditioners; they literally warm the indoors by cooling the outside. With mechanical energy (usually provided by electric motors), they extract and upgrade energy from the outside atmosphere, pumping it up to satisfactory indoor temperatures. A heat pump may yield 2–3 Btu's in heat for every Btu it requires in electric power. The heat pump cools when operated in reverse. From an energy standpoint, heat pumps appear to be an attractive proposition but problems of consumer acceptance, reliability, cost, and servicing persist. Also, supplementary resistance heaters would be needed. Widespread adoption seems, therefore, to be some time off. Moreover, since heat pumps are economically sound only with a high load factor (i.e., both winter heating and summer cooling), this innovation could raise summer air conditioning loads above levels that would otherwise prevail in the area.

Enhanced performance of room air conditioners could produce an important amount of energy saving and, equally significant, could cushion the impact of peak electric power demand in the New York downstate area. (Residential air conditioning accounts for about 40 percent of summer peak demand in the Con Edison service area.) At the present time, there is a substantial range of variability in the performance of given-sized window units—a variability for the most part not reflected in cost differences.

The New York State Ad Hoc Committee on Appliance and Apparatus Efficiency has concluded that substantial improvements in room air conditioner operating efficiency are "reasonable and well within the capability of present technology." Accordingly, the committee recommends the legal imposition of progressively higher minimum standards of performance, which (depending on size and voltage), for the period 1975–79, would mean improvements in "energy efficiency ratings" (EER) ranging from 30 to 40 percent.[38] EER's are ratios of cooling produced to power consumed, or Btu's per hour to wattage. In 1973, some representative 115-volt room air conditioning units on the market had the range of EER's shown on the following page.

Clearly, performance does improve with size of unit; the low end of

---

[38] *Report on Appliance and Apparatus Efficiency*, pp. 247–50.

Cooling capacity (Btu's/ hour)	EER range
4,000	4.7– 7.0
6,000	4.9– 8.8
8,000	5.7– 9.8
10,000	7.2–11.6

the range for 10,000 Btu/hour units exceeds the high end for the 4,000 Btu size. Still, variability at given sizes is great. The average EER for *all* window units appeared to be around 7.0. A mandated minimum averaging over 9.0 is deemed easily attainable by 1979.

Finally, a saving of approximately 5 percent of the gas consumption of new residential gas furnaces (as well as hot water heaters) could be achieved by replacing gas pilot lights with automatic ignition devices. Additional heating fuel savings could be brought about by reduced thermostat settings. A 68° F, rather than a 72° F, temperature setting could reduce the energy consumption of residential heating systems by an estimated 12 percent; at 65° F, the saving is figured at about 15 percent.[39]

Transportation. In the New York region, as elsewhere, the principal key to achieving energy savings in transportation involves one or more of the following changes:

1. More efficient performance of a given transport mode—e.g., an increase in miles per gallon for the average car on the road.

2. An enhanced load factor, as in the case of more car pooling for commuters or a higher proportion of occupied seats in aircraft.

3. Shifts towards more energy-conserving transport modes, illustrated by a switch from passenger cars to mass transit or from truck to rail freight.

One could add possibilities that are not simply adjustments internal to the fuel- or power-driven transport sector. Walking or limiting the number of automobile shopping trips saves fuel. More distant opportunities include reliance on closed-circuit videophone systems instead of traveling; or changed land use concepts and practices having the effect of reducing transport fuel consumption. In other words, the possibility

[39] Ibid., p. xvi.

Table 3-3.   Energy Use Per Unit of Travel, New York Region, 1970

	Passenger travel[a]		Freight transport[b]	
Travel mode	Passenger miles traveled per 100,000 Btu of fuel consumed	Percent of gross energy used in passenger travel	Ton-miles per 100,000 Btu of fuel consumed	Percent of gross energy used in freight transport
Bus	43.2	0.8	—	—
Rail	38.4	1.8	143.9	9.4
Subway	32.0	3.0	—	—
Auto	15.4	69.6	—	—
Airline	8.8	22.2	—	—
Taxi	6.7	1.9	—	—
Truck	—	—	12.9	78.6
Total	14.8	100.0	[c]	100.0

Source: Adapted from *REC*, p. 15.

[a] Not shown separately, but accounting for less than 1 percent of fuels used in passenger travel are: trolleys, ferries, and general (i.e., non-airline) aviation.

[b] Waterborne freight and goods shipped in air freighters are not shown separately. These account for about 12 percent of energy consumption in freight transport.

[c] Not calculable because of incomplete data.

of exogenous reductions in transport energy demand should also be kept in mind.

The first column of Table 3-3 records the well-known variability in energy efficiency for various passenger and freight transport modes in the New York region. Listed in descending order of efficiency use, the figures show that the most prevalent modes are well down in the efficiency rankings. Even though there are some problems of estimation and interpretation,[40] these data do dramatize the major points at issue in transportation energy use.

The preeminent share of regional energy consumption accounted for by automobiles has one clear-cut implication: even a minor improvement in average automotive fuel efficiency signifies major potential savings in transportation energy use. A recent analysis concerning energy use and transportation in New York State points out that a "4% saving

[40] For example, a tabulation for just the immediate New York City environs would reflect more efficient overall usage because of the greater relative weight of railways and subways. In addition there is some distortion arising from the inclusion of long-distance air traffic originating or terminating in the region's major domestic and international airports. Finally, one cannot make rash critical judgments about the ostensibly suboptimal rail-truck freight mix, since a limited region of the country offers less scope for intercity rail freight than the nation as a whole.

in automobile fuel consumption would be nearly as productive as a huge 50% reduction in fuel used by railroads within the state."[41]

The realization of improved gas mileage, coupled with better load factors so as to yield higher passenger miles per gallon of gasoline, would thus seem to rate as a prime objective of a transportation conservation strategy. Discouragement of ownership of large automobiles as well as improved engine performance of given-sized cars are obvious ways to go. A speculative example of fuel saved by an assumed shift toward smaller cars alone has been worked out in the study just quoted and is shown in Table 3-4. Under the circumstances conjectured, a decade's shift (sufficient time for turnover of most cars initially in existence) towards compacts and subcompacts could achieve a yearly gasoline saving of 10 percent below the consumption level otherwise projected. The associated weighted statewide miles per gallon would rise from 15 to 17. A gasoline saving of 489 million gallons—equal to 11.6 million barrels or about 32 thousand barrels per day—would result. With gasoline consumption accounting for about one-fifth of statewide energy consumption, this would signify an approximate 2 percent drop in New York State energy consumption, assuming other consumption practices were unchanged.

To the extent that even rough quantification is possible, other conservation potentials for automotive gasoline appear to be much more modest than a shift in size of cars.[42] For 1971 transportation patterns and levels of automotive fuel consumption, it is estimated that savings of some 35 million gallons/year of gasoline would ensue from pooling. Voluntary commuter auto pools, from all indications, appear to have low appeal where driving and parking conditions are thought generally adequate. Incentives and responses are therefore key questions about which little can be said. A 9 percent enlistment in auto pooling (beyond what already exists) has been described as an optimistic upper limit in New York State. No attempt has been made to hypothesize future developments.

[41] *Report of the Ad Hoc Committee on Energy Efficiency in Transportation* to the Interdepartmental Fuel and Energy Committee of the State of New York (Albany, N.Y., October, 1973), p. 17. (This is a counterpart study to the *Report on Apparatus and Appliance Efficiency*, cited earlier.)

[42] The three examples that follow all come from the *Report on Energy Efficiency in Transportation*. For further useful observations, see Michael Gerrard, *Transportation Policy and the New York Environment*, Council on the Environment of New York City (June 1974).

Table 3-4.   Fuel Consumption by Car Size, Historical and Projected, New York State, 1971 and 1980

Car category[a]	Miles per gallon	1971 Percent of cars in category	1971 Fuel consumed (million gallons)	1980 estimate Trend[b] Percent of cars in category	1980 estimate Trend[b] Fuel consumed (million gallons)	1980 estimate Shift toward small cars[c] Percent of cars in category	1980 estimate Shift toward small cars[c] Fuel consumed (million gallons)
Sub-compact	22	12.8	343	15	523	35	1,194
Compact	16	25.4	963	19	906	25	1,172
Intermediate	15	27.0	1,097	30	1,508	25	1,256
Standard	13	34.7	1,622	35	2,040	15	866
All cars[d]	15	100.0	4,032	100	4,986	100	4,497

Source: Efficiency in Transportation, pp. A3-2 and A3-3.

[a] The weight range (in pounds) for the four categories are, respectively: 1,000–2,499; 2,500–3,249; 3,250–3,749; and 3,750–5,249.

[b] As extrapolated by the New York State Department of Motor Vehicles.

[c] The "high" of two cases calculated in the source (above).

[d] A negligible quantity of special category cars, not shown separately, is included in the total. The total number of cars was 6.07 million in 1971 and is projected in both 1980 cases at 7.51 million. Average miles per gallon would remain at a weighted figure of 15 in the first 1980 case and improve to 17 in the second case.

Radial tires are believed to save from 5 to 10 percent of the fuel in highway driving, but less in urban driving since their use would not prevent braking energy losses. An average 3 percent saving seems reasonable. Assuming that 30 percent of New York State vehicle owners could be induced to adopt radial tires (again referring to 1971 experience), an estimated annual gasoline saving of 45 million gallons results.

An average 2 percent improvement in mileage has been ascribed to optimal (as opposed to prevailing) tune-up practices. When this improvement factor is applied to the state's 1971 car and truck population, a 20 million gallon/year saving is indicated. All told, the 100 million gallons/year savings attributed to pooling, radials, and tune-ups would have represented from 2 to 3 percent of 1971 New York State gasoline consumption, thereby reducing overall statewide energy use by another 0.2 to 0.3 percent (in addition to the 2 percent ascribable to the shift in car size).

A number of additional energy saving transportation potentials are even less susceptible to numerical estimation. One such proposal involves substantial improvement in the scale and quality of public transit service with a view to lessening people's dependence on private automobiles. Depending on the time-scale of implementation, a significant diversion might ultimately necessitate substantial government financing. The expansion of high-speed intraregional rail service is a related idea.

There is no doubt that changes in land use planning and control over new subdivisions could materially affect levels of demand for automotive travel. More than other parts of the United States, the New York region, and particularly New York City and its surroundings, already reflects an urbanized settlement pattern that is not rampantly dissipative of transport energy. More to the point, proposals such as those designed to produce indirect benefits in one area (energy savings) from practices exercised in another (land use) must weigh carefully any measured or unmeasurable costs likely to be incurred in the course of change. Crowding, noise, and pollution come to mind as conceivable offsets to energy savings in this example. We make no judgment as to the net balance of advantage but point only to the hazards of one-sided decision processes.

Finally, a word on freight transportation. In the New York region, the passenger–freight energy split is 79 percent and 21 percent, respectively. Although total transport energy usage is thus dominated by energy consumed in passenger transport, the freight portion is heavily tied to a transport mode—trucking—that is characterized by disproportionately large energy needs. (See Table 3-3.) But the New York State report quoted earlier sees only limited opportunity for realizing significant energy savings in this sector.[43] Given the limited size of the economic and marketing territory being examined (i.e., ruling out extensive rail shipments of, say, wheat, coal, or ores) major recourse to trucking seems almost inescapable. Some improvements, however, seem possible through the elimination of "gateway" city requirements in truck routing; and through governmentally aided efforts to help preserve branch rail lines, which, under present uneconomical circumstances, seem ripe for abandonment.

Other Energy Use Categories. Aside from the key energy use categories of space conditioning and transportation, several other areas of energy consumption are candidates for more economical consumption practices. We shall briefly mention a few of these.

[43] *Report on Energy Efficiency in Transportation,* p. 45.

Lighting ranks relatively far down the list of key energy uses—in New York State accounting for about 8 percent of residential electricity consumption[44] but undoubtedly for a considerably higher proportion of commercial sector usage. Nonetheless, it would appear to offer the potential for markedly more economical utilization. (At the very least, it would be better if the waste heat that illumination necessarily generates could be put to profitable use.) Lighting levels recommended by illumination engineers have been tending to rise over the years, but many questions are beginning to be raised concerning the physiological basis for such standard setting. A recent New York energy study suggests that the sensitivity of psycho-physical experiments to determine lighting needs shows that very large errors have probably been made in the past in translating experimental data into lighting standards. "All in all, experimental data should lead us to lighting specifications more than 30% lower than the IES [Illuminating Engineering Society] specifications."[45]

In any case, it has been suggested that a number of steps which it would be feasible to adopt over a number of years (especially in new buildings) "could easily result in a one-third reduction of energy used for lighting."[46] These steps include, among others, the design of more efficient luminaires, the placement of decentralized control switches,[47] the optimization of window design for daylight, the use of high reflectance finishes, and the use of more efficient light sources. The last item reflects the wide prevailing disparity of lamp efficiency. For example, stated in lumens per watt, a 112-watt fluorescent lamp registers 81 compared with about 18 for a 100-watt incandescent bulb. The economic justification for effecting such energy saving changes seems favorable given current and prospective price relationships, but it does need to be more systematically explored. In highway use, high pressure sodium vapor lamps have been shown to minimize both energy consumption and annualized cost.[48] (It is scarcely ever noted in the literature, but, in terms of energy efficiency, lampshades are poorly designed.)

[44] *Report on Appliance and Apparatus Efficiency*, p. 18.
[45] Taussig and others, "Energy Conservation in High Density Areas," p. 39.
[46] Ibid., p. xi.
[47] The New York World Trade Center, presently precludes individually controlled room lighting. But this is a widespread situation, not limited either to public buildings or to New York City.
[48] Ibid., p. 99.

Refrigeration (including freezers) is responsible for about one-fourth of New York State's residential electric power use. More careful operation and improved maintenance would appear to be a "painless" route to achieving more sparing energy use in this area, but, as in many other instances, a clear-cut payoff to the owner must first be recognized. (In single-metered apartment buildings, for example, the incentive to economize on electricity use is absent.) On the manufacturing side, more effective insulation seems to be the most promising means of achieving energy savings; changes in the design of compressors and evaporator and condenser coils are more problematical in their economic and energy-use consequences.[49]

By focusing, as we have, on discrete aspects of energy usage (such as insulation or lighting), we run the risk of encouraging disjointed perspectives or approaches to what should, ideally, be viewed as modules of broadly interconnected systems. For example, discussions concerned with the desirability of upgraded insulation standards sometimes leave the impression that this can expeditiously be handled by, say, building code revisions or other measures addressed to specific elements of building construction. But the matter is not that simple. The problem is that of devising energy performance standards within a framework *which regards the entire building as a system*, as opposed to the imposition of separate and fragmented requirements. One can conceive of a pair of buildings, the first of which was effectively insulated but lacked an efficient heating plant, electrical system, and ventilation; while the second, though poorly insulated, might consume less net energy because of a highly efficient and well-maintained electrical, lighting, and mechanical system. This applies to both residential and nonresidential structures.

Or, consider as another example the concept of "total energy systems"—i.e., integrated systems which provide electricity by on-site generation, using current technology, such as gas turbines or diesel motors. Total energy systems make possible waste-heat utilization for space heating, cooling, and other purposes, thus reflecting high energy-conversion efficiency. Variations on the system exist (e.g., on-site steam generation may be supplemented by purchased utility electricity).

Research in this area must address various, often conflicting, goals,

---

[49] Ibid., pp. 109–10.

touching on raw energy conservation, economics, and environmental aspects. For example, a Consolidated Edison Company evaluation of sixteen alternative energy "packages" for a planned, large-scale commercial–residential project (Manhattan Landing in New York City) claims that a total energy system ranked very high by raw resource conservation criteria. From an economic standpoint, however, it did poorly, being well outranked by on-site steam generation with purchased electricity, while consideration of environmental factors yielded more or less arguable results, since questions of tall utility stack dispersion vs. localized generation are central to the debate. (Con Edison mentions objectionable traffic noise, exhausts, and nuisance associated with frequent fuel oil deliveries to the site—gas being assumed unavailable.)[50] Some specific points to be addressed are:

1. How sensitive are total energy system economies to the cost of standby utility power?

2. Does the use of fuel oil, rather than natural gas, seriously affect the environmental virtues of total energy systems?

3. What state–local institutional or legal barriers, if any, exist with respect to total energy systems (e.g., the jurisdictional authority of regulatory commissions or the possibly inhibiting role of building codes and zoning laws)?

4. Can the total energy concept be generalized, or are variations due to location, specific environmental factors, fuel-supply uncertainty, etc. such that no useful overall statement can be made, each proposed system having to be studied on its merits?

These cautionary remarks are not meant to predicate the realization of important energy savings on the completion of more research alone. They do, however, call attention to nagging and complex issues whose recognition adds a necessary and sobering caveat to some of the more extravagant, but simplistic, claims sometimes made for energy-conservation potentialities.

[50] Remarks by Bertram Schwartz, (then) Vice-President, System Planning and Fuel Supply, Consolidated Edison Co., at the "Building to Save Energy" Seminar, New York City, April 13, 1973.

Summary

In this chapter, we have surveyed promising energy-conserving practices in various sectors of the economy—nationally and regionally. Both because of their dominant quantitative importance in total energy consumption and because of attractive opportunities for more economical energy utilization, we emphasized two types of use in particular: space heating and cooling and transportation. In the former case, the potential energy and monetary savings resulting from improved insulation practices deserve clear recognition. Enhanced performance of room air conditioners could produce an important amount of energy saving and, equally significant, help blunt the impact on peak electric power demand in New York City. In the transportation sector, even a minor improvement in automotive fuel efficiency could signal major savings in energy use. A number of energy uses, ranking relatively low in terms of their aggregate importance, nonetheless seem to offer worthwhile energy savings possibilities. Household and commercial lighting falls into this category. The importance of an integrated systems approach to decisions on energy optimization in large buildings was underscored.

4.

Policies to Implement
Energy Conservation

WHAT ARE the principal means for reducing the level and growth of energy consumption? As noted at the beginning of the preceding chapter, one approach—that of deliberately curbing economic growth in general so as to induce dampened energy use—is best ruled out by assumption. Without going into its merits, we see no indications that either the NYR or the United States in general is yet prepared to go this route by conscious resolve. Another extreme approach involves the exercise of coercion over energy consumption by imposing outright ceilings on amount of use or prohibition of particular uses. But that degree of control also seems at odds both with the nature of the problem and the societal setting, although resort to drastic allocative measures during short-term crises is not ruled out (e.g., the winter of 1973–74); nor is public intervention for even longer periods ruled out where specific supply or environmental problems are deemed to exist.

What is left is a twofold set of instruments for influencing consumers' decisions on energy use. The first is through the operation of impersonal market forces. The second involves the use of purposeful policy tools, which, in turn, can be regarded as being of two kinds: those that simply facilitate more rational behavior in response to given market forces, and those that actually attempt to improve the allocative function of the marketplace—for example, by providing signals reflecting social (or external) as well as nominal costs of energy production and delivery systems.

The illustration of automotive travel brings these somewhat abstract categories into focus. Sharply rising gasoline prices (induced, say, by higher costs of imported or domestic crude oil) can be presumed to have some downward effect on gallonage demanded, whether it takes the form of retrenchment in, or shifts away from, automotive travel, use of smaller cars, or more car pooling. Yet a rational response of this kind to market price changes might never occur without better information enabling the consumer to work through changing gasoline costs, fuel economy, car size, and car price to a "best buy" calculation. At least, the decision is apt to be less sensible than it might be. Thus, it is likely that compulsory labeling of the comparative MPG performance of cars and their comparative annualized fuel costs may evoke a more informed decision (one that is more sensitive to higher energy costs) than would otherwise be the case. Here, there has been no encroachment on the market mechanism. Naturally, just as energy-conserving consumers will be economically (if not necessarily psychically) better off than others, so producers and sellers of lighter cars are apt to gain larger market shares—the more so if, in addition to better information, minimum performance standards are actually prescribed. In the latter case, to be sure, some degree of compulsion enters the picture, and the line of demarcation separating this group of actions from those designed to modify the price outcome becomes somewhat blurred.

Where the market price is itself judged to be an inadequate indicator of the broader social costs of production and consumption—perhaps because of automotive pollution, congestion, or, on a different level, the national security risk of large-scale petroleum imports—then measures to alter the purely privately determined market outcome may be in order. Automotive horsepower or fuel taxes are frequently proposed to serve that objective.

In what follows, we discuss each of these approaches to energy conservation in turn, with some attention to the question of whether the measures are more appropriate to national or subnational policy initiative. We will also remark upon the equity implications of energy conservation measures.

The Impact of Changing Market Conditions

A preeminent historical characteristic of energy use in the United States is the long-term decline of "real" energy prices. Throughout much of the

20th century, the price of fuels and power has either fallen in nominal terms or, even when rising, did so far less steeply than the general price level (see Table 4-1). Hence, apart from other determining factors, declining relative prices have constituted a strong and enduring stimulant towards progressive increases in energy use. This says nothing about whether the stimulus was induced by prices regulated at artificially low levels, as in the case of natural gas, or was in fact a consequence of market-determined costs and prices.

Even during the greater part of the period since 1960, real energy prices were relatively stable, particularly for natural gas, heavy fuel oil, and electricity. In the case of electricity, real prices began to

Table 4-1.   Selected Energy Components of the Consumer Price Index, 1952–1972

(Index Numbers: 1967 = 100)

Years[a]	Actual index numbers for December of year shown			Relative to total consumer price index	
	Total consumer price index	Electricity	Electricity and gas	Electricity	Electricity and gas
1952	80.0	93.6	83.5	117.0	104.4
1953	80.5	93.1	84.7	115.7	105.2
1954	80.1	94.7	86.3	118.2	107.7
1955	80.4	95.4	88.1	118.7	109.6
1956	82.7	95.5	88.6	115.5	107.1
1957	85.2	96.3	90.3	113.0	106.0
1958	86.7	97.6	93.5	112.6	107.8
1959	88.0	99.6	97.0	113.2	110.2
1960	89.3	100.0	99.3	112.0	111.2
1961	89.9	100.1	99.4	111.4	110.6
1962	91.0	100.3	99.6	110.2	109.5
1963	92.5	100.0	99.6	108.1	107.7
1964	93.6	99.3	99.8	106.1	106.6
1965	95.4	99.1	99.5	103.9	104.3
1966	98.6	99.2	99.4	100.6	100.8
1967	101.6	100.5	100.2	98.9	98.6
1968	106.4	101.0	101.4	94.9	95.3
1969	112.9	104.2	104.8	92.3	92.8
1970	119.0	109.9	110.7	92.3	93.0
1971	123.1	116.0	118.2	94.2	96.0
1972	127.3	120.2	122.5	94.4	96.2

Source: BLS data shown in Edison Electric Institute, *Statistical Yearbook of the Electric Utility Industry for 1972* (New York, 1973), p. 67.
[a] Referring to December of year shown.

increase selectively around 1967, and by the early 1970s were rising throughout much of the NYR's franchise areas. Distillate fuel oil prices also tended to rise towards the latter part of the 1960–70 decade, as did coal and gasoline at the pump. These trends operated at the national level as well as in the local setting with which this report is concerned.[1] Table 4-2, summarizing some key data for New York City, shows that for virtually every use, energy prices had risen less— and often substantially less as in the case of electric power—than the overall consumer price index. The substantial annual rate of increase in national and regional electricity use, noted in Chapter 1, surely bore some causal connection with this falling relative price trend. As the historical pattern reverses itself, with energy prices outpacing the general price index, at least some bending in the slope of the energy demand growth curve seems a strong likelihood. The question is: What is the probable extent of consumers' demand response to the real energy price increases that almost all informed persons judge as inevitable? It is insufficient and premature to assume that the significant slowdown in demand following the 1973–74 Arab oil crisis conclusively demonstrated the nature and extent of future demand dampening. A number of factors (other than price) affecting that situation may not recur: a relatively warm winter, the effect of widespread exhortation, physical controls, and actual and feared shortages, to name just a few.

The last few years have seen a rising interest in, and a considerable amount of econometric research on, the subject of price and income elasticity of demand for fuels and power, that is, the quantity of energy demanded in response to specified changes in price and income. Most studies that have so far ventured any tentative answers have dealt with the price question as it affects residential electricity use and automotive gasoline consumption. In the New York region, these two uses alone approached 30 percent of net energy use in 1970; combined they grew at an average annual rate of 5 percent during 1960–70. Because of substantial technical problems, very few efforts have been directed at determining price or income elasticity for energy in the aggregate.

[1] Note that we are discussing price *changes* rather than *levels*; for almost all energy products, the level of prices is substantially higher in the New York region than for the United States in general.

Table 4-2. Trends in Selected Retail Energy Prices, New York City, 1960–1973[a]

	Current dollars or cents			Indexes (1960 = 100)[a]		
	1960	1969	1973	1960	1969	1973
Residential monthly bill (per customer)[b]						
Nonsubstitutable electricity[c]	$16.08	$21.08	$17.11	100.0	131.1	106.4
Substitutable electricity[d]	17.54	15.57	21.91	100.0	88.8	124.9
Substitutable gas[d]	9.41	9.06	10.44	100.0	96.3	111.0
Space heating						
Gas	27.16	25.04	30.37	100.0	92.2	111.8
Electricity	94.97	58.60	87.13	100.0	61.7	91.7
Oil	22.63	26.69	30.59	100.0	117.9	135.2
Air conditioning						
Gas	18.32	17.11	23.67	100.0	93.4	129.2
Electricity	16.77	18.98	24.20	100.0	113.2	144.3
Electricity price (per kwh) to large commercial users	1.97¢	2.50¢	3.49¢	100.0	126.9	177.2
Price (per gallon) of regular grade gasoline	33.9¢	32.9¢	41.9¢	100.0	97.1	123.6
BLS consumer price index:						
New York City[e]	—	—	—	100.0	124.5	156.9
United States	—	—	—	100.0	123.7	148.0

Sources: All energy data from Foster Associates, Inc., *Energy Prices of 1960–73* (Cambridge: Ballinger Publishing Co. for the Energy Policy Project of the Ford Foundation, 1974), Chapter 3. NYC consumer price index from Bureau of Labor Statistics, *Handbook of Labor Statistics 1973* (Washington, 1974), pp. 307–08; and *Monthly Labor Review*, April 1974; U.S. consumer price index from *Economic Report of the President*, various issues.

[a] Data refer to January of the year in question, except for the NYC consumer price index, representing an average of calendar year data for 1959–60, 1968–69, and 1972–73 respectively.

[b] For each category shown, the monthly bill is based on specified volumes of energy, which are constant over time (e.g., 400 kwh/month for nonsubstitutable electricity).

[c] Covers such uses as lighting, refrigeration, and small appliances.

[d] Covers such (nonheating) uses as water heaters, ranges, and clothes dryers, in which either gas or electricity may be used.

[e] New York City and northeastern New Jersey.

A recent survey[2] of existing elasticity studies discloses a considerable range in the resulting estimates. In the case of price elasticity of demand, the figures are:

Residential electricity	-0.8 to -1.3
Gasoline: short-run	-0.2 to -0.7
long-run[3]	-0.5 to -0.8

The limited work surveyed on total energy shows a price elasticity of demand of roughly from -0.2 to -0.5. What these figures mean— using, conservatively, the least elastic (most inelastic) end of the ranges—is that, all other things being equal, residential electricity demand would decline from levels otherwise prevailing by 0.8 percent following a 1 percent increase in its price (or, equivalently, by 8 percent following a 10 percent rise in price). A 10 percent hike in gasoline prices might, eventually, yield a 5 percent drop in demand; while a total energy demand decline of 2 percent might be associated with a 10 percent price rise in overall energy prices.

There are a number of reasons why such findings must be treated with a great deal of caution. (1) Studies based (necessarily) on a history of real price declines are applied to an environment of price increases. Symmetry may not prevail. (2) Relatedly, a paucity of historical time series has prompted investigators to rely upon a cross-sectional basis of analysis (where, for example, price and quantity consumed in state or utility franchise area A are compared with price–quantity relationships in area B). It is difficult to sort out price from other determinants of interregional energy use. (3) The substitutability of numerous energy forms (e.g., electric and gas heating) means that a sharp demand response to price change for a given energy form may merely shift some of that reduction onto increased demand for other forms. This, of course, explains why, as shown above, total energy demand is less price-elastic than individual energy forms. (4) Energy, being a derived demand in satisfaction of other goods and services (e.g., "capital" acquisitions such as stereos and cars), changes in energy

---

[2] C. M. Siddayao, "Estimates of Price and Income Elasticities of Demand for Energy: A Survey," a preliminary and unpublished draft report to the Energy Policy Project of the Ford Foundation, April 29, 1974.

[3] That is, allowing for an adjustment period to alter type of car ownership.

use will obviously react to changed factors affecting those other goods and services. (5) Elasticities are normally expressed as single values, but they may, in fact, vary at different price *levels*. Thus, a doubling of electricity prices at 2¢/kwh may trigger a less proportionate response than a 25 percent increase at 6¢. (6) It seems almost trite to add that a demand response to price presupposes an awareness of the energy cost being borne by the user, but where—as in parts of the New York region—single-metered apartment houses prevail, the electricity price probably has little direct effect on consumption habits. For this reason, multimetering is much to be preferred as a way of facilitating more informed energy-use behavior. (7) The demand restraint induced by price increases may be blunted to a greater or lesser extent by concurrent increases in income. The survey cited earlier shows the following income elasticities of demand:

Residential electricity	0.3 to 1.4
Gasoline	0.8 to 1.5

Thus, at the very least, a 10 percent rise in income (with prices unchanged) implies a 3 percent increase in residential electric power use and 8 percent in gasoline demand.

Of course, the economists who have developed these estimates are well aware of the various pitfalls of methodology, data reliability, and interpretation. We merely want to point out the considerable degree of uncertainty that still shrouds this relatively unexplored terrain of energy economics.

Even when the thrust of such elasticity studies is severely qualified (as is proper), it remains a legitimate presumption that the increasing real price for fuels and power that is in prospect over the next decade or so will exert some noticeable degree of restraint on consumption. If this is not the case, persons given to projecting a continuation of historical growth rates (e.g., 7 percent or more in the yearly growth of electricity consumption) are burdened with the argument that the drastic turnabout from falling to rising prices now occurring will have no appreciable effect. (To argue that income growth would neutralize the effect is illogical since it is the *break* with the past array of causal factors that concerns us.) A position envisaging undeflected energy growth would thus seem to be unsupportable. There is, however, a limit on the speed with which price-induced demand changes can take place. For where people are permanently or for many years locked into a piece of energy-using equipment (say, a particular home heating unit) or where a shift to more economical substitutes is precluded by nonavailability (say, the absence of adequate mass transit), a number of years may lapse before demand restraint begins to register.

Policies to Facilitate More Informed Energy-Use Decisions

Stimulating a Shift to Smaller Cars. Unquestionably, a progressive shift to smaller cars would represent a significant contribution to energy savings—a step, moreover, that seems, on balance, unlikely to have a socially disruptive impact of a serious sort.[4] Nationally, an average car weight of 2,500 lbs instead of the prevailing 3,500 lbs has been estimated as saving 2 million barrels of oil daily, which equates to nearly one-eighth of present consumption, or the amount of oil we expect to receive later this decade through the trans-Alaska pipeline.

Mandatory posted information on gas mileage and on yearly fuel costs (at different assumed gasoline prices) might effectively supplement higher prices as a spur to such a shift. A federally encouraged voluntary miles-per-gallon labeling program is now in effect. A strengthened, compulsory effort seems desirable and has been proposed as federal legislation. States and localities could independently enact such measures without inviting the sorts of counterproductive responses from other areas or inequitable assumption of burdens that might accompany other kinds of regional initiatives.

Encouraging Improved Building Insulation. We have already referred to the critical importance of building insulation in determining the energy needed for space conditioning, which is a ranking component of the NYR's energy use. The imposition of specific compulsory insulation standards is discussed in the next section. Here, we point merely to the desirability of compelling the disclosure to prospective homeowners of their annualized lifetime heating-and-cooling costs, at any given degree

---

[4] In fairness, we ought to mention the debate over whether smaller cars provide less protection to occupants or whether slower speeds and a *widespread* reduction of car size would cancel out a tendency to more serious accidents.

of insulation and type of space-conditioning equipment. Reasonable assumptions would also have to be made about intensity of use and estimated range of energy prices. It seems a fair presumption that, as prospective homeowners recognize how quickly they can recoup the initial capital cost through reduced operating expenditures, the housing market will be forced to respond to this new element in consumer choice.[5]

Of course, perceptible progress in this regard will be slow in much of the New York region since new housing represents a very small percentage of the existing housing stock. Table 2-3 shows that, for recent years, the Greater NYC Area's housing increment was about 1 percent of existing dwelling units. Moreover, apartment housing dominates much of the NYR. For example, the 1970 proportion of apartments in all dwelling units in NYC was: Manhattan—98 percent; Bronx—85; Brooklyn—69; Queens—50; and Richmond—19.[6] This suggests that only limited consumer leverage can be exerted through the policies proposed above. Mandatory insulation requirements will be needed, and even they will take a long time to make themselves felt.

Air Conditioning Efficiency. We noted earlier the significant (40 percent) contribution made by air conditioning to summer peak demand in downstate New York. Since over 60 percent of dwelling units in the Greater NYC Area lacked any kind of air conditioning as recently as 1970,[7] energy demand for this purpose can be expected to continue its substantial growth, thereby adding to the pressure on strained electric generating capacity. The New York State government task force de-

---

[5] What may be an attractive proposition for homeowners has been instituted by the Consumers Power Co. of Michigan, with the approval of the state's Public Service Commission. The company has started a home insulation service for its residential gas customers under which cost of insulation can be spread over 36 monthly installments (at 12 percent interest per annum) payable as part of the regular utility bill. With visible evidence that the insulation is paying for itself through lowered gas consumption, the program seems like an ingenious approach to at least this one facet of energy conservation. The company offers free insulation counseling and will arrange for installation, if requested.

[6] Census data shown in Brookhaven National Laboratory and State University of New York (Stony Brook), *Residential Energy Consumption and Income: A Methodology for Energy Policy Analysis Applied to the Greater New York City Region*, BNL 18818 (April 1974), p. 7.

[7] Census data shown in *Regional Energy Consumption*, Second Interim Report of a Joint Study by Regional Plan Association, Inc. (RPA) and Resources for the Future, Inc. (RPA, New York, N.Y., 1974), Table 23.

scribed earlier, recognizing the "unnecessarily" wide range of cooling performance (relative to wattage used) of room air conditioners now on
the market, has recommended legislatively mandated minimum standards
of efficiency.[8] It might be, however, that a mere mandatory *labeling*
requirement (certifying the efficiency and perhaps also the estimated
operating costs) would impel manufacturers to respond competitively to
this new factor in merchandising once it is forced on them. A mandatory
labeling requirement for room air conditioners was, in fact, enacted by
the New York City Council in 1973. The experience under that law
should be studied as to impact and effectiveness.

The Role of Education. Though the lasting quantitative impact is
debatable, a sustained effort by state and local government agencies in
the NYR to disseminate to local organizations and the public at large
information about well-conceived forms of energy conservation seems
well worth the small price it would probably cost.

State and local governments, in turn, should be systematically brought
into a framework of federally developed intelligence on energy conservation. The national government has the budgetary capability to
conduct and sponsor propmising energy R&D activities on a scale infeasi-
ble subnationally. The Department of Housing and Urban Development
and the National Bureau of Standards, for example, are deeply involved
in questions of energy use as a function of total building design and
particular energy systems.

One desirable consequence of efforts to mobilize comprehensive
information on energy conservation is that it will enable state and local
government units to subject their own expenditure decisions to the test
of economically rational energy use. Builders and contractors on such
public projects as schools, hospitals, and office buildings could be
induced to give an economic accounting of their proposed plans as they
relate to energy use. This should apply especially to insulation practices, space-conditioning equipment, and lighting. The same considerations might govern the acquisition of publicly owned vehicles. This is
not an argument to have public expenditure decisions subjected, as has
occasionally been proposed, to "energy impact statement," for the mere

---

[8] *Report of the Ad Hoc Committee on Appliance and Apparatus Efficiency*
for the Interdepartmental Fuel and Energy Committee of the State of New York,
Albany, New York, June 25, 1973, p. 33.

demonstration of Btu savings is an insufficient basis for decision making. Rather, the economic consequences of such savings and what it takes to achieve them deserve explicit airing.

It would be desirable if owners of residential and commercial property could have access to accurate physical and economic information (say, from engineering or architectural experts) on costs and benefits associated with changes in insulation, in space-conditioning systems, or in other energy use practices. But recourse to the information involves in itself a cost–benefit decision, which, without some degree of public sponsorship—at least initially—is apt to be taken only haltingly.

Ultimately, general educational efforts no matter how well-intentioned are unlikely to equal in impact the more specific actions discussed under separate headings in this chapter.

Policies to Improve Functioning of the Market

Stimulating a Shift to Smaller Cars. The automobile imposes extensive burdens on society, which, in many cases, are inadequately reflected in charges to users. True, gasoline taxes and tolls designed to cover highway construction, road maintenance, and related services do represent partial offsets to such costs. But when we consider air pollution, downtown congestion, the value of lost time suffered by blameless victims of such congestion, and the value of public land that has to be bid away from other uses to the accommodation of automotive traffic, a system of costing that is faithful to such "externalities" is largely absent. A horsepower or weight tax could help sway owners towards smaller cars, but it would have to be imposed at the federal level, since such a tax at the regional level would be frustrated if buyers could shop in nontaxing jurisdictions. However, future modification of state vehicle registration-fee schedules to encourage popularity of smaller automobiles does seem feasible. In New York State, such fees, based on gross vehicle weight, are already reasonably correlated with fuel consumption; perhaps the progressive "bite" in the schedule is too weak to be an effective deterrent to large-car purchases, but the point deserves study. The logic of such taxes would be enhanced if their proceeds could be deployed for plans to encourage more efficient forms of transportation.

Increased motor fuel taxes, assessed in cents per gallon pumped, would also constitute a disincentive on automotive energy use. For a number of reasons, however, it is a measure inappropriate to state, let alone local, units of government. First, purchasers could patronize lower-taxing jurisdictions nearby. Second, a substantial tax could generate bootleg trade in gasoline, which would either produce unequal impacts or would necessitate policing. More broadly, however, the use of state tax systems for shaping major social policy objectives, rather than producing revenue for support of direct expenditure programs, flies in the face of state fiscal-policy tradition and would probably encounter formidable public opposition. Even at the federal level, the likely regressivity of such a tax would spark intense controversy, unless accompanied by rebate provisions.

Measures for Other Automotive Energy Savings. Several other policy measures could contribute to more economical fuel consumption in the automotive market. To encourage a shift towards energy-saving radial tires, an excise tax on ordinary tires has sometimes been talked about. If such a tax were indeed appropriate on a regional or local level, it should probably be phased in gradually so as to avoid undue penalties on existing car owners, but it could become progressively steeper as the present stock of nonradials reach the end of their useful lives. Attempted evasion by the purchase of tires in nontaxing jurisdictions could be controlled by including the requirement for radials as part of vehicle inspection systems. But once again, federal taxing authority seems like the more practicable instrument of control. One might expect that, if radials are supposed to pay for themselves, that fact alone should be sufficient to prompt consumers to buy them. However, consumers often buy short-lived goods or buy on credit because the initial purchase price is generally much more apparent and decisive in purchasing decisions than long-term, hidden costs, especially when there is limited cash on hand. As with the payoff from insulation, many consumers discount the future more heavily than others and are willing to take short-run gains for long-term losses.

Mandatory tune-ups at specified frequency—a requirement initiated under the impetus of emission-control objectives—would also result in fuel conservation. Using reasonable assumptions about response and compliance, the New York State study group frequently cited above[9] ascribed very modest fuel saving potentials to these two measures (radials plus tune-ups)—on the order of ½ of 1 percent of New York statewide transportation fuel used.

Local regulations prohibiting, sharply curtailing, or discouraging automotive activity in central business districts are no doubt implementable through measures ranging from outright bans to steep parking fees or commuter taxes. Though aimed primarily at diminishing air pollution, as in the case of numerous transportation control plans under the Clean Air Act, such policies could also dampen overall fuel use. But, except for seriously polluted and congested areas (like Manhattan's central business district), such proposals should be treated with great circumspection. Restrictions on automotive use in the absence of alternative transport modes might merely accelerate a diversion of economic activity from city centers to suburbs—a process which, in the end, might actually cause a net increase in energy use, quite apart from the unsettling effect it could have on employment and business conditions.

Encouragement of Public Transit. The expansion of existing rapid transit services or the development of new systems are probably more glamorous than realistic options.[10] At least that is the case where really large savings of energy are envisaged. Rail systems involve huge capital costs, take a long time to build, and, where they eliminate bus systems having a more dispersed route structure, may on balance increase the passenger volume less than anticipated. What may be more promising in the New York region (and particularly in the Greater NYC Area), with its extensive existing system of rapid transit, are additional ways to expand the offpeak use of capacity which is intensively utilized during peak commuter hours. Lower fares for the elderly in offpeak hours, as well as half-fares on Sundays, are examples of promotional devices already in use in New York City.

The expansion of bus transit systems—less burdened by the enormous capital commitment of rapid rail service—deserves earnest policy

[9] *Report of the Ad Hoc Committee on Energy Efficiency in Transportation* to the Interdepartmental Fuel and Energy Committee of the State of New York (Albany, N.Y., October, 1973), pp. 35–36.

[10] See testimony by John R. Meyer, *Economic Impact of Petroleum Shortages*, Hearings, Subcommittee on International Economics, Joint Economic Committee, 93d Cong. 2d sess., 1974, pp. 107–14.

consideration. Special bus lanes on approaches to tunnels feeding into Manhattan (and similar devices on highways in the Virginia suburbs of Washington) illustrate the value of according buses priority use on existing streets and expressways. Present car drivers, however, would have to be shown the adequacy of bus capacity, speed, scheduling, and route dispersion; for, unless price differentials are huge, quality of service may be a more potent factor than price in wooing people from cars to buses. The advantage is that the availability of such service need not await either the rebuilding of cities or a massive infusion of capital. Even so, some considerable amount of both local and federal funding to meet some of the planning, operating, and capital costs seems indicated. The payoff both for energy savings and urban amenities would, however, be positive.

Freight Transportation. Policies designed to shift NYR freight transportation towards more energy-efficient modes—essentially, from trucks to railroads—would appear to be an act of futility. (Freight accounts for about one-fifth of the NYR's transportation fuel consumption, largely concentrated in trucking.) Railroads are too financially strapped to maintain economical branch lines and would require state and/or federal financial assistance to attract a significant share of the freight market. In the meantime, both federal and regional governmental regulations have to a slight degree enhanced trucking efficiency by permitting greater flexibility in routing requirements.

Encouraging Improved Building Insulation. The New York State Ad Hoc Committee on Appliances and Apparatus has called for mandatory statewide minimum standards of building insulation in new residential and commercial structures.[11] Pending the promulgation of permanent standards, interim insulation standards for New York State not lower than those prescribed by HUD for federally assisted housing are recommended by the committee. In addition, a determination of how much upgrading is feasible in *existing* housing is suggested.

Limited steps toward implementing these standards have already been taken. Under a New York State Public Service Commission (PSC) order, all new homes completed after January 1, 1975, must meet speci-

[11] See *Report on Appliance and Apparatus Efficiency,* p. ix.

fied minimum insulation standards in order to receive gas heating service. The standards include 6-inch insulation under the roof and in floors over exposed areas; 3.5-inches in sidewalls and over crawl spaces; and double glazing of windows and doors. The PSC, as early as 1971, concluded that the state's utilities could not meet unconstrained demand for natural gas and therefore imposed limits on new connections.

Whether the informational requirements (discussed in the earlier part of this chapter) and the establishment of minimum standards would go about as far as policy could in bringing about ultimately significant energy savings in space-conditioning, or whether still tighter regulations are needed, are questions which it would be premature to answer categorically. Certainly, there are policy measures which would seem desirable in the interest of a still more effective response—but such measures require study rather than quick implementation. As examples, two sets of questions, designed to unravel the traditional overriding concern with minimization of initial building costs, would constitute useful areas of research.[12]

1. Suppose the size of the required down payment in home financing came to be determined, in part, by projected annual operating costs— repayment of principal, interest, insurance, and other expenditures, *but explicitly including expenditures for fuels and power*. Loan repayment ability rises as operating costs decline. Assume, for example, that the mortgage for a $50,000 home required a 20 percent or $10,000 down payment. If the purchaser were willing to invest, say, an additional $5,000 (bringing the price of the house to $55,000) in insulating materials whose cost would be more than recouped in demonstrably lower energy bills, why subject the purchaser to the 20 percent "penalty" on the low-risk $5,000 investment?[13] The feasibility of this proposal merits study, but some questions should be addressed: By what cooperative or legislative means can financial institutions and builders be induced to accept this idea? How should the discounted value of future fuel and power prices be calculated? Is such a policy most appropriately coupled with the enactment of insulation standards? Since such

[12] See Joel Darmstadter and Eric Hirst, "Energy-Conservation Research Needs," in Hans Landsberg, and others, *Energy and the Social Sciences—An Examination of Research Needs*, RFF Working Paper EN-3 (Baltimore: Johns Hopkins University Press for Resources for the Future, 1974), pp. 407–09.

[13] Laurence Moss, testimony before U.S. Senate Committee on Interior and Insular Affairs, 93rd Congress, 1st sess., March 23, 1973.

incentive devices are aimed at, and only practicable in, *new* housing, some attention ought to be given to the possibility of including conservation practices in existing housing. Can present homeowners be realistically expected to pay for improved insulation now (through "retrofitting") even given the long-run economic advantage of doing so? The question is important, for even if one assumes that *all* new structures conformed to advanced insulation standards, the resulting reduction in the NYR's energy growth rate would be less than that obtained by the application of improved standards to just a small fraction of existing structures.

2. A study is needed to determine the effect of the tax system on energy conservation. For example, ownership of apartments—which were shown above to figure so importantly within the NYR—is completely depreciable in a short period of time, at which point it may become advantageous to sell. If the first buyer knows he is going to own buildings for only a brief interval, can he be induced to worry about the economics of energy conservation, or is he going to be primarily interested in saving capital? Study here could embrace commercial as well as residential structures.

Regulatory Aspects of Energy Pricing. There is one respect, finally, in which the influence of public policy upon the functioning of energy markets has received, and has deserved, prominent attention. That is the case where pricing for fuels and power—and, hence, demand–supply responses—departs from that which normally characterizes reasonably competitive markets. On the national scene, for example, many (though by no means all) competent observers believe that the manner in which the Federal Power Commission has exercised its responsibility to control the price paid by interstate pipelines to natural gas producers has led to a distortion of the public-interest objectives which the policy was designed to serve. By setting regulated natural gas prices at a point judged considerably below the "market-clearing" level, the commission has provided an artificial stimulus to demand (i.e., a disincentive to conserve), and, at the same time, failed to encourage development of new supply. At the very least, this has contributed to the scarcity of natural gas that now prevails in many parts of the nation, including the New York region. To the extent that certain tax benefits accorded the crude-oil producing industry (principally percentage depletion and intangible drilling expensing) involve a burden shifted from the price mechanism and the individual consumer onto the general taxpayer, a similar encroachment on the efficacy of market processes results. Note that this point does not depend on whether one supports or opposes oil depletion allowances. It merely says that if it is desirable for government to spur secure domestic supply, then "the cost of customer security should be paid by customers . . ."[14] rather than through devious and indirect instruments, such as the depletion allowances. Again, as in the natural gas case (though for different reasons), the price should not be providing misleading stimulus to demand above levels implied by a "true" market price.

In the context of energy conservation, the area where the appropriateness of publicly regulated energy prices has evoked perhaps most debate is the electric utility rate structures decreed by state regulatory bodies. At issue is the question of electric utility pricing and promotional practices, particularly that practice which provides for a progressively lower unit price (cents per kwh) as the volume of monthly kwh's consumed rises. In mid-1971, for example, Con Edison's monthly electric utility tariff for residential customers was as follows:[15]

First 10 kwh or less	$1.70
Next 50 kwh	4.80¢ per kwh
Next 60 kwh	3.80¢ per kwh
Next 120 kwh	3.05¢ per kwh
Over 240 kwh	2.45¢ per kwh

Since these so-called "declining block rates" appear to encourage increased electricity consumption at a time when capacity expansion plans are frequently frustrated, voltage reductions are common, and environmentally acceptable ways of generating power are limited, they have encountered a considerable degree of public criticism. The more extreme critics call for an opposite restructuring of electricity rates ("rate inversion"), under which unit prices would actually *rise* with

[14] Gerard M. Brannon, *Energy Taxes and Subsidies* (Cambridge: Ballinger for Energy Policy Project of the Ford Foundation, 1974), p. 74.

[15] Shown in Federal Power Commission, *National Electric Rate Book* (Washington, January 1972). Gas utility rates and electricity rates to commercial and industrial users follow a similar pattern. The actual level of unit prices in each size group has risen substantially since 1971.

greater use. More moderate proposals involve a flattening of the rate schedule whereby unit rates would be uniform across the different volume classes. The New York State Public Service Commission, in a 1971 rate decision, forced Con Edison to raise its electricity rates to large industrial users by a greater percentage than those to small industrial users. More recent decisions have moved in the same direction. Both events suggest that the traditional rate structure may now be subjected to some degree of regulatory reform.

The traditional approach to utility rate-making practices arises from several considerations: First, progressively lower per-kilowatt-hour electricity costs are defended on grounds of economies of scale. That is, a lot of electricity can allegedly be produced at a cheaper unit cost than a little electricity. Second, the bulk of the cost of transmitting and distributing power, a considerable part of fixed distribution plant (such as substations and transformers), and even—to some extent—generating capacity are needed to service customers irrespective of how much electricity they consume. In other words, costs for the foregoing items as well as for meter installation and billing do not vary with volume of demand and therefore represent a charge per unit of electricity sold which declines as volume increases. (In the Con Edison schedule, this fixed charge is reflected in the $1.70 item shown in the first line.) Thus, the first factor, reinforced by the second, leads to falling unit costs of electricity as consumption rises.

In principle, the system of electric utility rate-making considered most efficient by economists is marginal cost pricing: that is, where rates cover the incremental costs of the last additional unit of electric output as well as the marginal cost of capacity needed to accommodate the long-term growth in consumption. Such marginal cost pricing is regarded as resulting in a socially optimal allocation of resources. Some critics of declining block rates argue that, whatever their past justification, prevailing rates in the large volume blocks are not justified by economies of scale and are below marginal costs, as a result of which consumers are given incorrect price signals and encouraged in "uneconomic" consumption. Thus, a recent EPA report[16] illustrates how instituting a constant marginal electricity rate at various levels of demand (i.e., a "flattened" rate schedule) need not jeopardize near-term utility reve-

[16] M. R. Seidel, S. E. Plotkin, and R. O. Reck, *Energy Conservation Strategies*, U.S. Environmental Protection Agency (Washington, July 1973), pp. 42–43.

nues; though, to the extent that such rate reform induces dampened demand growth, utility revenue growth would ultimately be affected as well. Incidentally, since a fixed customer charge would apply irrespective of the quantity of electricity supplied, the average kwh charge per user (customer charge plus constant marginal electricity cost) would still exhibit a falling *average* total cost per kwh the greater the consumption of electricity.

Quite apart from proposed fundamental reforms such as the preceding, there is one important and particular respect in which cost-of-service precepts are frequently ignored in utility pricing, and that is in the case of peak load power use. There is evidence that peak power may be much more costly to produce than off-peak output, yet it is usually sold at no greater price, particularly in residential use. In NYC, for example, the peak summer demands occasioned largely by the concurrence of air conditioning, rush hour subway transport, and early evening household use has meant the costly deployment of gas turbine generators for meeting the extra load.[17] Shift towards a dual pricing system would therefore seem to be an economically justified approach which would serve to blunt one of the major problems in energy use, even if—by inducing people to divert some usage from peak to off-peak times of the day—the totality of fuels and power consumed might not appreciably change. Peak and off-peak pricing, as well as even more sophisticated approaches to rate design, exist in other countries, such as France. Initially, the cost of dual-meter installation would have to be faced—and economically justified—but this cost would be a legitimate charge in the interest of a presumably more rational and beneficial utility pricing system. But even where peak power is priced—as it should be—at its true cost, we cannot be sure of consumer response under extreme conditions of need, such as on the hottest day of the summer. For at that point, it may be quite difficult to blunt demand, so that the power system's capacity must be designed to meet that very contingency or to allow load-shedding as a normal practice. If people try to offset their peak surcharge by cutting back at off-peak, the consequence could actually be to reduce the utility's load factor and to raise overall costs —an unintended result, but one which apparently occurred in a number

---

[17] Note, however, that in the Con Edison area, shortage of generating capacity has dictated the installation of costly gas turbines even for base-load electric power production.

of places in the United States during the 1973–74 conservation campaign.

A less drastic approach than dual metering involves seasonal peak surcharges, which a number of utility systems now impose during the summer air conditioning months. But even this more modest proposal is regarded with disfavor by some regulatory commissions. For example, in a recent rate increase for the Public Service Electric and Gas Company of New Jersey, the New Jersey Public Utilities Commission reportedly rejected the recommendations for seasonal peak pricing made by the state's Department of Environmental Protection, Energy Office, and Economic Council.[18]

Whether more drastic rate restructuring would be justified in the interest of deliberately curtailing power demand raises much more intricate issues. To begin with, one would have to demonstrate that the declining rate is clearly inconsistent with rational pricing policy. There is dispute about this, perhaps not so much among economists as among people in the utility industry. Then, too, under rate inversion, one would want to explore whether demand restraint induced by higher rates to large-volume users might not simply offset increased demand in the low-volume categories. All this depends on the respective elasticities, about which little precise understanding exists. One would also have to know whether, and to what extent, a price-induced suppression of electric power demand diverts demand to substitute energy forms. In the case of large-volume industrial users, purposeful non-cost-related price increases to restrain demand might spur moves to more accommodating geographic areas. There hangs over all this the question of whether the utility rate structure, whatever its inherent defects, is a legitimate vehicle for achieving broad social objectives which are substantially external to it, though misgivings in that direction must be tempered by the possibility that the structure now existing may obstruct such objectives rather than be neutral.

Equity Questions

By and large, we have been stressing economic efficiency criteria as the sensible overall approach to energy conservation. There is no point

[18] *The Journal of Commerce,* September 18, 1974, pp. 3, 9.

in spending a dime to chase a penny. To the extent that prices in the marketplace properly reflect, or by deliberate policy are *made* to reflect costs, resource scarcity, and environmental damage, the selection of energy-using processes or activities which save Btu's but result in higher costs leads to deterioration, not enhancement, of aggregate social welfare. For in such cases, a single resource would be conserved at the expense of other scarcer and more valuable resources.

But it is fair to go on and ask whether ground rules appropriate to overall standards of economic efficiency also conform to reasonable goals of social justice. Specifically, what are the interconnections between energy use, on the one hand, and consumers in different income or need categories on the other. In spite of demonstrable inefficiencies in energy use (some of which we have cataloged) and forgetting about such highly publicized (but statistically trivial) frivolities as electric toothbrushes, the fact remains that the principal components of energy consumption—notably heating, cooling, and transportation—involve activities most people would view as essential in their daily lives. No less essential for being less direct is the energy needed in business, farms, and factories in support of income and employment.

The essentiality with which households regard fuels and power in their budget is illustrated in Table 4-3. The table shows the estimated percentage of family income expended by different income groups in New York City and its suburbs for electricity and for heating fuels. Although the total household energy budget is not covered here (e.g., transportation is excluded) and the table is only a rough approximation, the figures are nonetheless instructive. They point to the extent to which the relative burden of energy outlays is very much greater in lower- than higher-income groups. This means that, in the absence of differential price elasticities of demand, a given increase in energy costs obviously results in a greater incremental burden (in terms of income) for low- than high-income groups. In other words, energy price increases have a regressive impact, much as in the case of food. If, as seems likely, price elasticity is higher at high-income levels (where there is room to shift to substitutes and to cut out "nonessentials") than at low-income levels, this impact is compounded.

It is the regressive phenomenon which, in the case of electric power, has sometimes led to proposals for cushioning the impact of energy price increases by the imposition of differential rate hikes, varying by

Table 4-3.   Expenditures for Electric Energy and Heating Fuels Relative to Income, New York City and Suburbs, 1970

Income group ($)	Electric energy[a]		Heating fuels[b]	
	Average annual bill ($)	Percent of family income	Average annual bill ($)	Percent of family income
		New York City		
Under 2,000	105	10.5 ⎫	117	7.8
2,000–2,999	105	3.5 ⎭		
3,000–4,999	108	2.7	120	3.0
5,000–6,999	112	1.9	123	2.1
7,000–9,999	117	1.4	128	1.5
10,000–14,999	124	1.0	139	1.1
15,000–24,999	135	0.7	147	0.7
25,000 and over	155	0.4	151	0.4
		Suburbs[c]		
Under 2,000	120	12.0 ⎫	178	11.9
2,000–2,999	120	4.0 ⎭		
3,000–4,999	128	3.2	185	4.6
5,000–6,999	129	2.2	191	3.2
7,000–9,999	132	1.6	216	2.5
10,000–14,999	146	1.2	231	1.8
15,000–24,999	156	0.8	259	1.3
25,000 and over	180	0.5	290	0.8

Source: Brookhaven-SUNY, *Residential Energy Consumption and Income*, pp. 39, 40, 47, 50.

[a] Estimated indirectly (from Census data on appliance ownership, utility rate schedules, and other data) rather than from information reported by occupants.

[b] Space and water heating use. Covers homes with distillate fuel oil use only.

[c] The LILCO service area in the case of electric energy; Westchester, Nassau and Suffolk counties in the case of heating fuels.

consumption level. However, quite apart from the appropriateness of such a policy from an economic standpoint, in New York City such a move would not appear a workable strategy. This is because the absolute amount of electricity consumption varies relatively little as between such contrasting income groups as the $2,000–$3,000 bracket, on the one hand, and the $7,000–$10,000 class, on the other. This characteristic probably arises largely from the high incidence of apartment dwellers with a lower and limited scope for appliance ownership.

Note from Table 4-3 that for the same income group, suburban outlays for heating fuels are substantially higher than in the city, due no doubt to the preponderance of single family dwellings. Any cost increases for heating fuels will thus hit suburban families harder than the corresponding income group in the city. One should not, however, stretch conjecture from very limited statistical evidence. A grossly underheated city dwelling—and there is little doubt that these exist to a worrisome degree—can scarcely be described as protected against sharply rising fuel costs by the mere fact of its low level of fuel usage.

The loudest howls of "foul play" don't always reflect the greatest inequities. Probably the most outraged outcry about higher energy costs in recent years came from Con Edison electric heating customers who, in just one year, 1973–74, saw their monthly electric bill rise by between two-thirds and 100 percent to the extraordinary level of around $250. Whatever their shock and grievance, these people elected (or had foisted on them) the costliest heating mode, received the preferential tariff of large-volume customers, and certainly weren't in the very low income brackets. That fact, however, did not deter the New York State Public Service Commission, in February 1975, from ordering temporary rate relief to downstate electric-heating customers. The PSC directed the affected utilities to recover the cost of the heating-rate reduction by raising the rates to all other customers;[19] a case of economic justice *twice* tempered by regulatory mercy.

Equity issues also arise in the transportation field. Higher gasoline prices or taxes, whatever the justification, would generally hit the poor harder than the well-to-do.[20] Also, to a large, low-income family barely able to own *one* car (which therefore may be one of ample size), a restructuring of the motor-vehicle fee schedule discouraging large cars may seem unreasonable. In one somewhat peculiar respect, the regressiveness of higher gasoline prices may, in effect, be blunted. Families owning *no* car at all and therefore far more restricted in their travel would no longer be as inequitably positioned, particularly if higher gasoline prices or taxes would focus more attention on the provision of satisfactory public transport.[21]

Short-term supply limitations are also likely to take a disproportionate toll among low-income groups. The long service station lines during

[19] State of New York Public Service Commission, News Release, February 7, 1975.

[20] The degree of severity would probably vary with the geographic area—more severe, say, in the suburbs than in New York City, where car ownership is held down by the presence of public transport; but we have not explored that question.

[21] A colleague points to an implementable gasoline "tax" which is thoroughly progressive: disallowing the deduction for gasoline taxes on both federal and state income tax returns.

the winter of 1973–74 were surely hardest on the one-car family bread-winner, whose vigil in the queue had to be carved out of the work schedule.

One is pressed to prescribe approaches to the equity problem that don't sound lame. At the height of the 1973–74 crisis, some proposals for a stiff tax to curtail gasoline demand also included the notion of a rebate or tax credit allowance for low-income persons. There may be occasions when the severity of an energy price increase warrants some such mechanism. At the very least, public intervention in energy markets —even when addressed to legitimate economic efficiency objectives— ought to be very keenly evaluated as to any social welfare implications. Perhaps educational programs relating to more economical energy use ought to contain a special effort directed to the interests of lower-income groups, though—as in the case of income maintenance pro-grams—poor people might argue that home economics advice can be a pretty flimsy substitute for a significant improvement in living stand-ards. And in the broadest sense, the energy crisis is one of both demand and supply; it may well be that a comprehensive energy program deal-ing with a greater *supply* capability can make as important a contribu-tion in blunting hardships of poor people as efforts in conservation. The problem is that few if any supply strategies seem to promise relief in the short run.

Summary

In evaluating measures that would lead to the adoption of energy-con-serving practices reviewed in the preceding chapter, we distinguished among the following instruments for influencing consumers' decisions on energy use: The operation of impersonal market forces; tools to facilitate more rational responses to given market forces; and tools to shape market forces so as to enhance their allocative performance.

We stressed that evolving market conditions—for example, the almost inexorable rise in real electricity prices that we are now witnessing—will inevitably induce a degree of retrenchment in use that would not have been indicated from past trends and that will occur independently of deliberate policy actions. But in order to enable consumers to respond knowledgeably to market conditions, policies designed to guide consumer behavior along a more informed path are clearly desirable. Examples include mandatory information on energy efficiency and costs in the heating and cooling of newly constructed buildings, the operation of automobiles, or in the use of room air conditioners.

In our third category, policies designed to affect the allocative outcome of market forces are brought into play. For example, a federal horsepower or weight tax would help sway owners towards smaller cars. The expansion of public transport—particularly bus transit systems, which are less burdened by the enormous capital commitment of rapid rail service—deserves earnest policy consideration. In housing, compulsory insulation standards and some changes in home financing arrangements favoring energy conservation practices suggest themselves. We discussed the matter of utility rate structure as it relates to the allegedly artificial encouragement towards electric power use. Although this issue is complex and perhaps not quickly soluble, progress in one area seems achievable. There is evidence that peak power may be much more costly to produce than off-peak output, yet is usually sold at no greater price. Shift towards a dual peak and off-peak pricing system would therefore seem to be an economically justified approach which would serve to cushion at least one of the major problems in energy use.

Finally, we raised the question of how energy-conservation ground rules appropriate to overall standards of economic efficiency may or may not conform to reasonable goals of social justice. Data on household budgets point to the regressive nature of fuel and power price increases—a fact which confronts policy makers with a dilemma, attracted as they are to the demand restraint induced by high prices, yet sensitive to budget ceilings of low-income people. At various points in the chapter we commented upon the probability that state and local governments are apt to play a less effectual role in energy conservation policy than the federal government.

5.

Growth in Energy Consumption
With and Without Conservation

THE ARRAY of energy-conserving practices and policies which we have surveyed in the last two chapters does not translate into a unique quantifiable specification of dampened energy-demand growth in the New York Region in the years ahead. Rather, we have sought, by using examples of selected potential payoffs, to depict some of the principal technical, economic, and policy circumstances that are pertinent to energy-conservation decisions. If this survey lacks the conciseness of a mathematical model, it is hoped that what remains is nevertheless more than a set of vague perceptions. With this in mind, it is useful to try and pin some suggestive numbers on the impact of feasible conservation efforts on energy-demand growth in the thirty-one-county New York region. This is done in Table 5-1.

The table reintroduces historical data on the NYR's energy growth, as well as projections of what future growth might look like without deliberate energy-conserving actions and measures. This is termed the trend projection. By contrast, the dampened growth scenario shown in the table assumes modest downward deflections in trend of a magnitude deemed nondisruptive, sustainable, and well-founded economically. Terms like "modest," "nondisruptive," etc. are, of course, value-laden and, to a considerable extent, judgmental. The following paragraphs may, however, help to dispel the excessively subjective tone of our formulation.

Residential Consumption

In dealing with the electric energy portion of the residential sector, recall that total NYR money income (in real terms) is projected as growing by 4.3 percent yearly (Table 2-1). Let us be cautious and say that residential electricity demand would increase proportionately, even though fragmentary statistical analysis suggests an income elasticity of demand below unity. On the extremely conservative assumption of some increase in real electricity prices and *some* price elasticity of demand (however low), residential electricity demand would go up by less than 4.3 percent per annum. However, let us again be conservative and project it as high as 5 percent per annum. Among other things, this would allow (a) for a time lag in consumer response to higher prices, and (b) for the contingency of electricity having to assume part of the burden of deficient natural gas supplies.

The nonelectric component of residential energy use is dominated by space-conditioning, which, for convenience, we can use as a crude proxy for the entire nonelectric share. The trend estimate for 1985 shows consumption of 1.27 quadrillion Btu. Of the NYR's 8.9 million dwelling units projected to exist in 1985 (Table 2-2), an estimated 1.7 million (roughly 20 percent) will have been constructed between 1974–85. Let us assume that this 20 percent of the 1985 housing stock incorporates energy-conserving practices yielding a 20 percent reduction in fuel use below what is implied in the trend line. A postulated reduction of this magnitude, applied, as it is, to *future* housing units, seems not unreasonable in the light of our earlier discussion of insulation and other payoffs. Arbitrarily, a 5 percent reduction in use in pre-existing dwelling units is assumed. Again, this figure appears quite realizable given the likely upward pressures of energy prices and the recognition of economically feasible conserving actions.

The Commercial Sector and Public Facilities

Between 20 and 25 percent of the commercial floor space projected to 1985 (see Table 2-3) is ascribable to construction activity occurring after 1974. It is this portion of 1985 energy use in the commercial/public segment that is most susceptible to the adoption of more economic energy use practices. We have applied a 15 percent reduction in fuel and power use to this incremental portion of 1985 structures and a very modest 4 percent reduction to pre-existing buildings.

Industry

This sector accounts for a relatively small share of the NYR's energy economy. The region's industrial energy requirements are met indirectly: it is in Pittsburgh steel plants, Detroit automobile factories, and chemical complexes in the Southwest that energy gets embodied into many manufactured products imported into the region. For that reason, as well as because the energy implications of technology and production processes in industry cannot be generalized, we have given little attention to the possibilities in that sector. In any case, industrial energy-use decisions can be assumed to respond closely to market forces. Numerous industry studies have suggested energy economies of around 15 percent below prevailing usage. Allowing for a "phasing-in" interval, we apply a 10 percent factor to the 1985 trend projection in electric power use, but little reduction in nonelectric uses whose growth is barely positive to begin with.[1]

Transportation

The projected trend level in 1985 electricity demand (essentially subways and electrified railways) in this sector is fairly small; we have no basis for a further reduction. Indeed, within a context of energy-conservation policies, some rise may be indicated as an offset to shifts away from other forms of transport. But that prospect is recognized only symbolically—by a slight increase over the trend estimate.[2]

The major component in the nonelectric transport category, and the

---

[1] Having observed some impressive and innovative energy-saving practices in industry in the last couple of years, a reviewer of the manuscript deems the 10 percent factor to be quite a bit too low.

[2] The capture of heat loss in subway braking is a promising and potentially significant energy saver beyond 1985. On the other hand, the cooling of underground stations and introduction of higher performance equipment for faster acceleration could work in the opposite direction.

one which we have addressed in particular, is, of course, automotive travel (with a 56 percent share in 1970). Unlike the case with housing, a decade into the future permits turnover of most of the equipment on the road in the mid-1970s, so that the 1985 trend figure represents usage dominated by transport equipment yet to be produced. A variety of factors can, and probably will, impel economies in automotive fuel usage, relative to the trend projection: rising gasoline prices or taxes, some decline in average car size, improved engine performance, enhanced load factors, and at least some momentum toward public transport. (Note that these items are not advanced as being additive; i.e., the price disincentive cuts across the range of factors mentioned.) We have applied a 12 percent fuel saving to the projected 58 percent of the 1985 transport sector accounted for by automotive travel. This figure is substantially less than reductions commonly thought achievable. Some manufacturers achieved such improvement during 1974–75 alone, and in his message to Congress, October 8, 1974, President Ford set a goal of as much as a 40 percent improvement in automotive gas economy on new cars over a five-year period. On the other hand, laudable as they are, these targets do involve a substantial measure of uncertainty. For example, can they be met in the face of future emission-control and safety standards?

The nonautomotive portion of the projected fuel demand in transportation offers less promise of rapid adaptation to new economic circumstances and to exploitation of new technologies and potentials in transport. (Remember, also, that at least some part of the automotive fuel savings will show up in a demand on other transport modes.) But here, too, some net improvement in energy efficiency is surely likely. We have gauged the possible savings at 5 percent.

Total Energy Savings and Some Implications

Table 5-1 shows the overall quantitative impact of the foregoing hypothetical elements designed to scale down the trend in energy consumption growth. Under the dampened growth assumptions, the level of electric power consumption in 1985 falls to about 15 percent below the projected trend case. Net energy consumption in the aggregate—comprising both fuels and direct power use—is about 10 percent lower. The same relative reduction occurs with gross energy consumption, which includes the consumption of fuels used to generate electric power. The biggest relative reduction in the projected percentage growth rate occurs, correspondingly, in electric power whose trend rate of increase (6 percent per annum) falls to 4.7 percent. Net energy consumption is dampened from 3.1 to 2.4 percent; gross energy consumption from 3.5 to 2.7 percent.

These figures, if they materialize, would reflect a conspicuous, if not sensational, momentum toward deceleration in the growth of energy use. They are certainly a far cry from zero energy growth. And they are compatible with further advances in per capita energy consumption. (Population growth is projected at under 1 percent yearly.)

Yet, distinct benefits could flow from the dampened growth we have plotted out here. For example, lower electric power consumption translates into a 1985 saving of 33 million Mwh or, say, 5,000 Mw of installed capacity. This could reduce by four or five the twenty-three to twenty-five additional big power plants in the NYR otherwise indicated for the period to 1985 (compared with around fifty existing in the early 1970s). For the Greater New York City Area alone, it might mean the equivalent of two fewer, large-sized plants. Environmental problems associated with energy use—especially power-plant emissions and automotive combustion—would, of course, be mitigated to at least a moderate extent. And the "raw" energy savings (748 trillion Btu) shown in the last line of Table 5-1 translate to around 380 thousand barrels/day in oil equivalent—with no guarantee, of course, that any such savings would not be dissipated by consumption outside the region.

Of course, retrenchment in electricity use during 1974, steeply rising prices, and a deeper appreciation by the utilities of the dynamics of demand may now be in the process of making the trend projection an even more hypothetical standard of reference than we have conceded it to be. To that extent, fewer "savings" are implied.

Whatever the case, our exercise obviously does not constitute a definitive brief for going slow on energy supply expansion. If anything, the *nearer-term* electricity supply needs seem pressing. What does seem called for is a very careful, dispassionate look at expanding electric power needs in which the risks of being caught short can be weighed against those of overexpansion. Utility forecasting in the past often failed to acknowledge even the possibility that the future growth curve may involve

Table 5-1. Energy Consumption, New York Region, 1960 and 1970, and Projections to 1985 Under Two Trend Assumptions

| | Energy Consumption (trillion Btu) | | | | Average annual percentage rate of change | | | |
| | | | 1985 projection | | | 1970–85 | |
	1960	1970	Trend[a]	Dampened growth	1960–70	Trend[a]	Dampened growth
Residential							
Electricity	43.0	98.0	287.3	204	8.6	7.4	5.0
Nonelectric energy use	901.4	973.2	1,268.8	1,167	0.8	1.8	1.2
Total	944.4	1,071.3	1,556.1	1,371	1.3	2.5	1.7
Commercial and public facilities							
Electricity	52.0	112.7	264.1	246	8.0	5.8	5.3
Nonelectric energy use	610.0	757.5	1,062.1	990	2.2	2.3	1.8
Total	661.9	870.2	1,326.2	1,236	2.8	2.8	2.4
Industrial							
Electricity	40.9	71.6	135.1	122	5.8	4.3	3.6
Nonelectric energy use	213.0	244.3	259.0	255	1.4	0.4	0.3
Total	254.0	315.9	394.1	377	2.2	1.5	1.2
Transportation							
Electricity	9.9	10.6	13.0	14	0.7	1.4	1.9
Nonelectric energy use	717.0	1,129.9	2,074.6	1,886	4.6	4.1	3.5
Total	726.9	1,140.6	2,087.6	1,900	4.6	4.1	3.5
Net energy							
Electricity	145.8	293.1	699.5	586	7.2	6.0	4.7
Nonelectric energy use	2,441.4	3,104.9	4,664.5	4,298	2.4	2.8	2.2
Total	2,587.2	3,398.0	5,364.0	4,884	2.8	3.1	2.4
Net energy excl. electricity	2,441.4	3,104.9	4,664.5	4,298	2.4	2.8	2.2
Utility fuel inputs[b]	546.2	1,080.9	2,349.5	1,968	7.1	5.3	4.1
Gross energy consumption	2,987.6	4,185.8	7,014.0	6,266	3.4	3.5	2.7

Sources: 1960 and 1970 data from Table 17, *REC*; 1985 "trend" data from Table 2–4; "dampened growth" data as discussed in accompanying text.

[a] "Trend" should not be construed to mean mechanical extrapolations of past developments. Reasonable assumptions about future tendencies in demographic and economic factors can yield departures from historic trends in energy consumption even before building in the consequences of assumed energy conservation practices in the "dampened growth" scenario. See the text discussion at beginning of Chapter 2.

[b] Fuel needed to generate electricity produced within region and imported from outside region. Also includes fuel loss in steam generation.

marked breaks with the past, irrespective of demonstrable changes in underlying conditions of growth. In utility-load forecasting, such myopia can have perniciously self-fulfilling consequences insofar as the capacity, once on line, imposes strong pressure on those obliged to promote its maximum use. It can also prompt the adoption of premature, and there-fore, uneconomical, options predicated to a large extent on high elec-tricity demand. But, as noted, recent events may be giving rise to a deeper probing of these issues.

Under any circumstances, it should be recognized, as a matter of near-arithmetic certainty, that a 10 percent-or-so savings in NYR fuel and power consumption in 1985 will be sopped up by ensuing growth within five years thereafter. That is, the consumption level of 1990 or even earlier will equate to the trend level in 1985. Whatever problems are associated with energy growth, they are postponed or eased at a point in time, but not resolved. Yet even this modest and—in our judg-ment—quite conservative deceleration in growth may generate a momentum towards a conservation consciousness which, once absorbed into the national value stream, can lead to a still sharper bending of the energy growth curve than we have hypothesized.

Summary

In this chapter, we have contrasted the trend projections of Chapter 2 with a scenario of dampened growth in the region's energy use. The latter projection—more of an illustrative exercise than a rigorous for-mulation—evolved from the analysis (in chapters 3 and 4) of feasible energy-conserving practices and policies. Under the dampened-growth assumptions, the region's electric power growth would come to 4.7 per-cent yearly, as against 6 percent in the trend projection. For net energy consumption in the aggregate, the respective figures are 2.4 and 3.1 percent. Under growth dampening, the level of regional electricity use in 1985 would be 16 percent below trend—savings which could signal a pronounced lessening of the strain on regional generating capacity; the overall level of net energy consumption would be 12 percent below trend. In the light of the opportunities which exist, these are modest savings, predicated on conservative assumptions. More dramatic con-servation potentials than those hypothesized would appear realizable, particularly as the time span for their adoption is lengthened.

6.

Concluding Remarks

TWO POINTS seem worth emphasizing or, since they have surfaced at one time or another in the course of our discussion, reemphasizing. The first concerns limits as to the effectiveness of purely regional energy conserving actions and policy initiatives. The second is the need to relate energy conservation to broader economic and social considerations.[1]

Regional success in energy conservation is much more likely if cast within a nationally directed and nationally adopted program than if policy initiative and response is limited to the state or locality. It is conceivable that such potentially high payoff measures as automotive horsepower taxes could be adopted by one taxing jurisdiction without a neighboring one doing so, but as a matter of political reality and, considering problems of border-crossing evasion, the prospects seem meager. Whether regionally mandated insulation standards higher than national ones would be effective depends on whether the traditional overriding concern with first costs could be overcome so as to preclude a shift of construction activity elsewhere. It is in the area of legislatively compelling greater disclosure as to insulating effectiveness, gasoline mileage, air conditioner efficiency, and so on, that regional policy can play an instrumental role in encouraging conservation. The New York City ordinance requiring efficiency labeling in room air conditioner sales is a small example. An innovative approach to utility rate making (such as peak-responsibility pricing) is another area where policies of state regulatory commissions need not await federal initiative. The New York State Public Service Commission seems to be far ahead of regulatory bodies elsewhere in the nation in at least deliberating on the appropriateness of such new policy departures.

---

[1] I am indebted to my colleague, Leonard Fischman, for some of the thoughts (conveyed orally and in writing), reflected in this section.

But even where regional energy conservation succeeds (irrespective of what triggered it), in the absence of a national commitment to conservation, savings in the New York region may simply be dissipated elsewhere by additional consumption of the freed resources. That being the case, the impulse to conserve on the part of the regional energy user may quickly flag. This happens because energy products move within a national, not a local, market. The exception is electric power capacity, where regional retrenchment in demand can confer visible benefits to the New York region, if only in the form of reduced claims on power-plant sites. In this respect, less regional consumption is beneficial even where the overall resource saving is cancelled.

Obviously where a long-term regional energy conservation strategy requires a large infusion of capital—as in the case of substantially expanded public transport facilities—a joint national–regional commitment would seem essential.

None of the foregoing is meant to apply to crisis circumstances, which are not touched upon in this study. In the context of an allocation program—where specified amounts of energy products are apportioned to states under national authority—policies at the state or local level to dampen overall fuel and power use or to prescribe priority uses are clearly a necessity.

Throughout these pages we have underscored the pitfalls of looking at energy conservation in isolation from a broader economic and social perspective. A good deal of the proliferating literature on energy conservation, however well-intentioned, is permeated with the single-minded goal of saving Btu's—short shrift being given to the trade-offs against which such savings ought to be reckoned. Distortion in this area is rampant, the examples ranging from the ludicrous (can automatic ventilating fans in windowless bathrooms be dispensed with; can stop-and-go driving be abated by replacing some stop signs with yield signs) to the simply misguided (because rapid rail transit is far less energy-intensive than automotive transport, it is the preferred travel mode, despite the cost of conversion). The lure of bicycling is rarely qualified by reference to substantial risks of bodily injury. A popular theme among energy conservationists is the inefficiency of energy conversion in food production: vastly more calories are embodied in the production process (e.g., in fertilizer) than are contained in the food yield—as if one had the option of eating natural gas instead. With no apologies for suburban sprawl, the energy "profligacy" that characterizes single-family housing (contrasted to the energy-saving pattern of more clustered development) is surely only one of a diverse number of resource, environmental, and amenity features that need to be taken into account in striking a decent balance in the area of housing and land use.

The urgency for dampening the growth rate in energy consumption is great; the opportunities for doing so in a meaningful way are numerous. It is hoped this report has helped identify some of these opportunities. What we need to guard against is the disposition to present the savings from conservation as if they were their own reward—cost free and pure gravy. In analyzing energy conservation, the watchwords should always be "costs and rewards," notwithstanding the fact that, in practice, it is unlikely that the valuation of either costs or benefits of conservation can ever be reduced to a specific dollar figure. More practically, the evaluation of energy conservation costs may suitably be viewed in terms of consumption values foregone and trade-offs for more costly nonenergy inputs. Indeed, in the interests of maximum analytical credibility, a conservative stance might be prudent: one might be well advised to start with the assumption that there are probably few *totally* costless conservation measures, since even such conservation as involves only more efficient means of achieving the same level of end-use consumption is apt to carry with it an investment either in new capital or in R&D, which diverts productive effort and personal income from the satisfaction of other current consumption. The economic costs of lost time are also relevant: enhanced efficiency of intercity buses compared with air travel (in terms both of fares and fuel requirements per passenger mile) may be washed out by the value of wasted time associated with the first transport mode.

Finally, what some energy conservation proponents label waste may be perceived by some individuals as a desirable service, well worth a premium in energy costs. In considering, say, frost-free refrigerators and self-cleaning ovens—both time and drudgery savers—we should resist judgment about energy adequacy. This is terminological sleight of hand, which ignores a legitimate diversity of choice and satisfaction even in such superficially homogeneous services.

This is not to take serious issue with the position that a questioning of some of the excesses of our affluent society and a possible toning down of our frenetic lifestyles may have civilizing rewards of their own. It does caution that, under circumstances where distinctions are murky at best, we avoid as much as possible a blurring of differences between, on the one hand, the value judgments of social critics and, on the other, real costs to individuals and society, which, if not directly measurable, can at least be identified.

Even where a healthy package of energy conserving practices is implemented, it is well to keep in mind that such steps are likely to make essential, but not decisive, inroads into the energy problems confronting the country. Thus, in the case of the New York region, even a fuller range of long-term conservation practices than we have hypothesized for 1985 would merely reduce the level of energy consumption to that which would have occurred around 1980. Alternatively, it would postpone for no more than a decade beyond 1985 the consumption levels that would have occurred in that year in the absence of conservation.

In its national implications, this would seem to suggest that efforts to cope with resource stringency and environmental problems depend on much more than just the kind of demand limiting actions we have been discussing in this report. But that clearly does not negate the desirability of implementing conservation efforts, which economic rationality should prompt us to do in any case. If the adoption of energy saving personal habits and commercial and industrial practices relieves us of (if only by postponing) some portion of the environmental burden, balance-of-payments difficulty, and resource pinch, thus giving us somewhat greater maneuverability in fashioning prudent long-range energy strategies, the undertaking will have been well worthwhile.

Conversion Factors[1]

As initially measured in the basic statistics, the variety of energy forms that aggregate to the NYR's net or gross energy consumption is expressed in such physical units as kilowatt hours of electricity, cubic feet of natural gas, or gallons of gasoline. In order to convert these physical terms into an additive and "common denominator" calorific units of measure (in our report, the Btu or British thermal unit), the table of equivalents shown below was used. (The Btu measures the amount of energy needed to heat one pound of water one degree Fahrenheit.)

Natural gas	1,030 Btu/cubic foot
Gasoline	125,000 Btu/gallon (= 5.2 mill. Btu/barrel)
Kerosene, diesel, jet fuel	135,000 Btu/gallon (= 5.7 mill. Btu/barrel)
Distillate (#2) fuel oil	138,700 Btu/gallon (= 5.8 mill. Btu/barrel)
Residual & bunker (#6) oil	149,700 Btu/gallon (= 6.3 mill. Btu/barrel)
Coal	25,000,000 Btu/ton
Steam	1,093 Btu/pound
Electricity	3,412 Btu/kwh

Where we refer to gross (as opposed to net) energy consumption (as in tables 1-1 or 2-4), we need to convert delivered electricity and steam not by their inherent thermal content but, rather, by the Btu equivalent of the fuels burned at power stations to generate electricity or produce steam. (In the electricity case, this figure is referred to as the heat rate.) For 1970, these conversions were:

Steam	1,588 Btu/pound
Electricity	12,372 Btu/kwh

These rates, which reflect the efficiency of conversion, change with time. In our report, we project slight improvements in the electric heat rate to 1985, but leave the steam conversion rate unchanged from the 1970 figure.

For those whose statistical orientation in the energy field runs to barrels per day of crude oil, one barrel per day equates to approximately 2,049 million Btu per year. Thus, regional gross energy consumption of 4,186 trillion Btu in 1970 (from Table 1-1) corresponds to a crude-oil equivalent measure of 2 million barrels per day.

[1] For further notes and data, see *Regional Energy Consumption*, pp. 19, 36, 37, 39, and 40.

Selected Bibliography

The following listing is, at once, more limited and broader than citations referenced in the study. Among sources already cited, only items judged to be of more general interest to persons concerned with energy conservation are included. That is, press releases, newspaper articles, narrowly focused pieces, and publications which have been superseded by more recent ones are omitted. Indeed, we recognize that many citations in the book (particularly those relating to conservation practices—in Chapter 3) should, if the study were initiated today, reflect more heavily the rapidly proliferating quantity of conservation and demand studies which have been appearing since the October 1973 Middle East war. For that reason, the bibliography is broadened to include some of these more notable recent references, even though they may not have been mentioned in the book.

The bibliography is divided into two sections: the first is applicable to the New York region specifically; the second, to energy conservation topics in general.

Studies Dealing with New York Region

Carolyn Harris Brancato and Jeffrey Cohen, *Electricity Demand and Supply in the Service Territory of Consolidated Edison Company of New York, Inc.,* New York: Citizens for Clean Air, Inc., Oct. 1973.

Brookhaven National Laboratory-State University of New York (Stony Brook), *The Effect of Specific Energy Uses on Air Pollutant Emission in New York City: 1970-1985,* BNL 19064. Upton and Stony Brook, N.Y.: Sept. 1974.

———. *Energy Supply and Demand in the New York City Region,* by H. G. Mike Jones, and others. (Unnumbered, preliminary printing). Upton and Stony Brook, N.Y.: Dec. 1974.

———. *Energy Use in the New York City Region,* BNL 18880. Upton and Stony Brook, N.Y.: May 1974.

———. *Residential Energy Consumption and Income: A Methodology for Energy Policy Analysis Applied to the Greater New York City Region,* BNL 18818. April 1974.

Council on the Environment of New York City, *Energy and the New York City Environment.* New York: 1974.

Gerrard, Michael. *Transportation Policy and the New York Environment.* Council on the Environment of New York City. New York: June 1974.

Patterns of Energy Consumption in the Greater New York City Area: A Statistical Compendium. RFF Working Paper EN-2. Washington: Resources for the Future, July 1973.

Regional Energy Consumption. RPA Bulletin 121. New York: Regional Plan Association, Jan. 1974.

Report of the Ad Hoc Committee on Appliance and Apparatus Efficiency to the Interdepartmental Fuel and Energy Committee of the State of New York. Albany: June 1973.

Report of the Ad Hoc Committee on Energy Efficiency in Large Buildings to the Interdepartmental Fuel and Energy Committee of the State of New York. Albany: March 7, 1973.

Report of the Ad Hoc Committee on Energy Efficiency in Transportation to the Interdepartmental Fuel and Energy Committee of the State of New York. Albany: Oct. 1973.

Sander, Diane E. *The Inverted Rate Structure—An Appraisal.* Part I—Residential Usage. Department of Public Service, State of New York, OER Report #IX. Albany: Feb. 17, 1972.

State of New York Public Service Commission, *Report on Energy Conservation in Space Conditioning* (Case 26292, Report by William K. Jones, Commissioner). Albany: Jan. 31, 1974.

Taussig, Robert T. and others. "Energy Conservation in High Density Areas: A Study of the Energy Crisis in New York City," Columbia University School of Engineering and Applied Science, Sept. 20, 1974 (draft manuscript).

Zupan, Jeffrey M. *The Distribution of Air Quality in the New York Region.* Baltimore: Johns Hopkins University Press for Resources for the Future, 1973.

Studies Dealing with Energy-Conservation Topics in General

Berg, Charles A. *Energy Conservation Through Effective Utilization.* National Bureau of Standards, NBSIR-73-102. Washington: Feb. 1973.

Chapman, Duane, and Timothy Mount, "Modeling Electricity Demand Growth," in Milton F. Searl, ed., *Energy Modeling,* RFF Working Paper EN-1. Washington: Resources for the Future, 1973.

Darmstadter, Joel, and Eric Hirst, "Energy-Conservation Research Needs," in H. H. Landsberg, and others, *Energy and the Social Sciences—An Examination of Research Needs.* RFF Working Paper EN-3. Baltimore: Johns Hopkins University Press for Resources for the Future, 1974.

Donnelly, Warren. *Energy Conservation Legislation in the 93d Congress.* Congressional Research Service, Library of Congress. Washington: July 22, 1974.

Gyftopolous, Elias P., Lazaros J. Lazaridis, and Thomas F. Widmer. *Potential Fuel Effectiveness in Industry.* A Report to the Ford Foundation Energy Policy Project. Cambridge, Mass.: Ballinger, 1974.

Hirst, Eric. *Energy Intensiveness of Passenger and Freight Transport Modes: 1950–1970.* Report ORNL-NSF-EP-44, Oak Ridge National Laboratory. Oak Ridge: April 1973.

——— and John C. Moyers. "Efficiency of Energy Use in the United States," *Science,* March 30, 1973.

Moyers, John C. *The Value of Thermal Insulation in Residential Construction: Economics and the Conservation of Energy.* Report ORNL-NSF-EP-9, Oak Ridge National Laboratory. Oak Ridge: Dec. 1971.

Seidel, M. R., S. E. Plotkin, and R. O. Reck. *Energy Conservation Strategies.* U.S. Environmental Protection Agency, Report EPA-RS-73-021. Washington: July 1973.

Perry, Harry. *Conservation of Energy.* Committee Print, U.S. Senate Interior and Insular Affairs Committee. 92d Cong. 2d sess. Washington: 1972.

Real Estate Research Corp. *The Costs of Sprawl.* Report Prepared for U.S. Council on Environmental Quality, Department of Housing and Urban Development, and Environmental Protection Agency. 2 Volumes. Washington: 1974.

Science Magazine, April 19, 1974. Entire issue on "Energy."

Sharefkin, Mark. *The Economic and Environmental Benefits from Improving Electrical Rate Structures.* U.S. Environmental Protection Agency, Report EPA-600/5-74-033. Washington: Nov. 1974.

Siddayao, Corazon M. "Estimates of Price and Income Elasticities of Demand for Energy." Unpublished report prepared for Ford Foundation Energy Policy Project, April 29, 1974.

Stanford Research Institute. *Patterns of Energy Consumption in the United States.* A Report to the U.S. Office of Science and Technology, Executive Office of the President. Washington: 1972.

A Time to Choose. Final Report by the Ford Foundation Energy Policy Project, Cambridge, Mass.: Ballinger, 1974.

U.S. Congress. House of Representatives. Joint Hearings, Subcommittees of the Committees on Government Operations and Science and Astronautics. *Conservation and Efficient Use of Energy.* 93d Cong. 1st sess. Washington: 1973. In 4 vols.

U.S. Congress. Senate Hearings, Committee on Commerce. *Energy Waste and Energy Efficiency in Industrial and Commercial Activities,* 93d Cong. 2d sess. Washington: 1974.

U.S. Department of the Interior. *United States Energy Fact Sheets by States and Regions.* Washington: Feb. 1973.

U.S. Department of Transportation and U.S. Environmental Protection Agency. *Potential for Motor Vehicle Fuel Economy Improvement.* Report to the Congress. Washington: Oct. 24, 1974.

U.S. Federal Energy Administration. *Project Independence Task Force Report—Energy Conservation.* Vol. 1, Residential and Commercial Energy Use Patterns 1970–1990; Vol. 2, Transportation Sectors; Vol. 3, Energy Conservation in the Manufacturing Sectors 1954–1990. Washington: Nov. 1974.

U.S. Federal Power Commission, *Practices and Standards: Opportunities for Energy Conservation.* National Power Survey, Report and Recommendations of the Task Force on Practices and Standards to the Technical Committee on Conservation of Energy. Washington: Dec. 1973.

U.S. Office of Emergency Preparedness, Executive Office of the President. *The Potential for Energy Conservation—A Staff Study.* Washington: Oct. 1972.

Wildhorn, S., B. K. Burright, J. H. Enns, and T. F. Kirkwood. *How to Save Gasoline: Public Policy Alternatives for the Automobile.* RAND Corporation, Report R-1560-NSF. Santa Monica, Calif.: Oct. 1974.

Index

Air conditioning, 18, 20–21, 22, 29, 46–47, 54, 56–57, 84; proposed standard for window units, 56–57, 74–75

Automobiles, 21, 31–32, 50, 58t, 60t; energy conservation policies, 66–67, 73, 76–77, 78–79; energy conservation potentials, 58–61; projection, 94

Car pooling, 59

Coal: pollution control regulations, 2, 15

Conservation: *see* Energy conservation

Consolidated Edison Co., 3–4, 17, 22, 82–83, 84n, 88

Conversion factors, 102

Cornwall pumped storage plant, 3–4

Climate, 14

Commercial floor space, 27t, 31, 92

Density, 29, 34, 40n, 61, 99–100

Economic growth, 25–31, 25t, 26t, 27t, 37, 41–42, 92

Economic welfare, 85–89, 99–101

Electricity, 2, 3–4, 17, 18, 55–56; consumption, by income classes, 87t, 88; metering, 63, 72, 84; New York Region compared with nation, 15, 15t; peak demand problem, 21–22, 40, 56, 74, 83–85, 98; power plant generating capacity, 37–40, 38t; prices, 67–73, 68t, 70t, 82–85, 92; projected growth rate, 29–32, 92, 96t, 94–97; utility expansion plans, 24, 95–96; utility rate structure, 82–85, 98

Energy: conservation (*see* Energy conservation); consumption and income distribution, 85–89, 87t; determinants of regional demand for, 9–10, 25–33; income elasticity of demand for, 72–73, 92; interrelationship of different types, 15; national consumption patterns, 17–19, 20t; New York Region consumption compared with nation, 10–11, 12t–13t, 15t, 16t, 18t, 20t; price elasticity of demand for, 69–72, 85, 86, 92; prices, 3, 19, 67–73, 68t, 70t, 81–89; projected regional growth of, 23–35, 25t, 26t, 27t, 28t, 30t, 91–97, 96t; regional consumption patterns, 10–17, 12t, 15t, 16t, 19–22; resource inputs, 18t, 15–17, 31, 34, 41, 97; supply issues, 2–3, 4; terminology, 13t, 26n, 26–27, 102

See also: specific types of energy

Energy conservation: commercial sector and public facilities, 47–48, 92–93, 96t; different concepts of, 36–42, 38t, 99–100; effect of market forces, 67–73; equity issues, 85–89; industrial sector, 49–50, 93, 96t; national vs. subnational government policies, 73, 75, 76, 77–78, 98–99; policies for achieving, discussed, 67–89; policies for achieving, summarized, 66–67; regional potentials, 51–64, 94–97, 96t; residential sector, 44–47, 92, 96t; space heating and cooling, 52–57, 73–75, 79–81; summary of studies on, 43t; transportation sector, 50–51, 57–61, 73, 93–94, 96t; transportation sector, utility rate structure, 82–85, 98

Environmental issues, 14, 37–39, 55, 63–64, 78, 94, 95

Freight transportation, 61, 79

Greater New York City area, defined, 4, 6f

Heat pump, 55–56

Housing, 26t, 74, 79–81, 92

Industrial energy consumption, 40–41; New York Region compared with nation, 10–11; projected growth in, 31

107